T0202371

LIBERATING SCIENCE

There are books that extol science and there are books that critique science, but there are few books that help the reader to see both the power of science and the constraints within which it operates. This perceptive and elegantly agued volume does what its title suggests; it helps liberate science from the misunderstandings that so often accompany it. *Liberating Science* can be read with profit by an intelligent sixth-former, the interested general reader and by specialists, all of whom will find much here that is insightful and thought-provoking.

Professor Michael J. Reiss,
Professor of Science Education, University College London

A beautiful exposition, by a prominent physicist, of the problems that underlie the way science is popularly presented by scientists themselves, and goes deeper into why wholly reductionist paradigms need to be relegated to history. It is clear, logically rigorous and highly accessible. A book that needs to be, and surely will be, widely read.

Dr Iain McGilchrist
Quondam Fellow of All Souls College, Oxford,
author, *The Master and His Emissary*

Not just engaging reading; it reaches a level of philosophical rigor that far exceeds what is usually on offer in discussions of science's relation to religious consciousness. I'm especially appreciative of the profound metaphors used to open up new conceptual space. Steane's writing also displays a due sense of humility about how each of us should think about our own response to the universe. Insightful, upbuilding, and potentially transformative.

Professor Hans Halvorson
Stuart Professor of Philosophy, Princeton University,
author, *The Logic in Philosophy of Science*, Cambridge University
Press, 2019

Andrew Steane has one of the University of Oxford's great scientific minds. That is no secret. Fewer people know that he is also a thinker of distinction in themes familiar to the arts and humanities. Wonder after wonder of science is presented here with marvellous clarity, in a case-study of how to avoid the reductionism that can plague popular science writing. Steane's science closes or reduces nothing: it opens up the world, and the human heart. His *Science and Humanity* taught me a good deal. More than that, it also moved me profoundly. *Liberating Science* has done the same.

Dr Andrew Davison
Starbridge Associate Professor in Theology and Natural
Sciences at the University of Cambridge

ANDREW STEANE

LIBERATING SCIENCE

The Early Universe, Evolution and
the Public Voice of Science

OXFORD
UNIVERSITY PRESS

OXFORD
UNIVERSITY PRESS

Great Clarendon Street, Oxford, OX2 6DP,
United Kingdom

Oxford University Press is a department of the University of Oxford.
It furthers the University's objective of excellence in research, scholarship,
and education by publishing worldwide. Oxford is a registered trade mark of
Oxford University Press in the UK and in certain other countries

Published in the United States of America by Oxford University Press
198 Madison Avenue, New York, NY 10016, United States of America

British Library Cataloguing in Publication Data
Data available

Library of Congress Control Number: 2023905962

ISBN 978–0–19–887855–1

DOI: 10.1093/oso/9780198878551.001.0001

Printed and bound by
CPI Group (UK) Ltd, Croydon, CR0 4YY

ACKNOWLEDGEMENTS

I would like to thank everyone who encouraged me in the writing of this book, and who helped improve the text. This includes anonymous referees who gave insightful feedback on a selection of chapters, and several people who kindly read the whole text. Michael Reiss, Michael McNamee and Iain McGilchrist all gave careful reactions and detailed suggestions. Such help is of inestimable value to an author. As always, the author remains responsible for whatever defects remain!

I acknowledge also the support of Oxford University in allowing time for me to learn and to write. This is not to claim any endorsement of any opinion expressed here, but rather to say that the aims of the University include aspects of what this book is about: the public understanding of science, the promotion of sound learning in physics and biology, and support for education in general. The academic values of the University, and the way they are handled in practice, have supported me.

I thank Henry MacKeith for informed and meticulous copy-editing which improved both this and an earlier book. In particular I acknowledge his thoughtful contribution to some of the poems, identifying instances where a more precise or apt word may be available.

I thank my son Joseph for taking the time to read and give feedback, and my family generally for bearing with me. Emma especially deserves all the thanks that I can give, for patience and perseverance, good nature and wisdom.

CONTENTS

CONTENTS

So his mind turned
to hall-building: he handed down orders
for men to work on a great mead-hall
meant to be a wonder of the world for ever;
it would be his throne-room and there he would dispense
his God-given goods to young and old—
but not the common land or people's lives.

Beowulf, Anglo Saxon, *c.*700–1000 CE,
trans. Seamus Heaney

CHAPTER 1

A CANDID FRIEND

Science has a fascinating and beautiful story to tell about the development of the universe at large and about the processes of life on Earth which led up to the world as it now is, and our place in it. Scientific writers have rightly wished to show this story to the general reader, both for the sheer pleasure of it, and for what wisdom it may furnish. This book is a contribution to that creative effort. So one major aim of the book is simply to open up areas of physics and biology for the interested reader. But in doing this, I want to achieve two further aims: first, to show that evolutionary biology has been misread by those who think it replaces moral philosophy; and second, to liberate science from private ownership. By 'private ownership' I mean the view that science belongs to an entirely naturalistic world-view and cannot otherwise be fully embraced. The book's title has a double meaning: a release of ourselves, and a release of science. In the first direction I want to show a reading of physics and biology which is not just correct but also, thankfully, more liberating than the one widely disseminated in recent years. In the second direction I want to encourage everyone to see science as part of our common inheritance, valued and able to be furthered by all of us, not the property of just one world-view calling itself 'scientific'.

Liberating Science. Andrew Steane, Oxford University Press. © Andrew Steane (2023).
DOI: 10.1093/oso/9780198878551.003.0001

To be clear, this is not to resist scientific methods and deductions; it is just to say that those methods are not able to grapple with everything we need to think about. They contribute, but they cannot fully elucidate issues of justice and value and purpose in the natural world, especially in human life. Science itself has to rely on values and assumptions. We can hold those values dear without thinking we obtained them by deduction from impersonal data, because we did not. They came to us, as a community, by other ways. I will say more about this as we go on.

The book approaches these aims mostly by presenting areas of physics and biology as accurately as I can while avoiding technical language. In particular, I (a physics professor of thirty years' professional experience) have done my best to identify tropes and assumptions in modern science writing which do not correctly convey what the research science is saying. Sometimes in the translation from technical terms to everyday language, ideas take root in the popular imagination that masquerade as 'what science says' but which really are not. So this book is in part a kind of cultural critique. It is an effort by a scholar at the University of Oxford, UK, to allow that university to exercise some intellectual leadership by pushing back against widely diffused misrepresentations of science. One of these misrepresentations came out of this same university, so this book also represents a bit of housekeeping.

The three misrepresentations I want to tackle are (1) the notion that scientific study is able to explain how there can be a physical universe as opposed to no physical universe; (2) the notion that biological evolution is without direction or purpose; and (3) the claim that social, aesthetic and moral experience is largely explained by evolutionary genetics. But my aim is not merely negative. It is also constructive, because it will be interesting to see how science does in fact operate in these areas, and because if we handle the presentation of science correctly then it will open up a more fair and open future for everyone.

In the chapters that follow I will examine these points in the order 1, 2, 3. But here I will introduce them 3, 2, 1.

The third point is that much of what is true concerning the nature of humans and other animals (and plants too, for that matter) is almost entirely independent of the processes central to evolutionary biology. This will push back against a wide swathe of contemporary culture which has swallowed the idea that the genetic mechanism dictates what the products of the mechanism will be. It is not so, just as the silicon microprocessor chips do not dictate what apps will be found on your mobile phone.

When we read, as has become all too common, that this or that facet of human behaviour is 'because' it tended to result in sexual attraction or tended to stimulate another's protective instincts or things like that, then the word 'because' is being used almost entirely incorrectly. In fact our behaviour is the way it is largely because that is the only way for us to navigate various physical, social and psychological situations, irrespective of whether anyone finds it attractive. When a man climbs a mountain he behaves differently to how he does when he swims a lake, and the differences in the motions he exhibits are not owing to evolutionary history or cultural pressure; they are owing to the fact that rock is solid and water is liquid. In a similar way, when people behave better when they are encouraged, and behave worse when they are threatened, this has almost nothing to do with the Darwinian process, and almost everything to do with the innate nature of relations between persons, no matter how those persons came to exist in the world.

In the chapters that follow I will recount this at greater length, and also some other points about evolutionary biology, such as the marvellous way in which it is shaped by the innate properties of the areas of life which are discovered by its process. This includes the areas of human life. Here is another illustration to place alongside the one about the climber and the swimmer. Observe that if we want to find out whether a mathematical idea such as Pythagoras' theorem is correct, then a study of our genetic history will not help. It will be irrelevant. We know this in relation to mathematical theorems. Something similar is true of insights in ethics and justice and subjects like that. The situation with social relationships is not quite as cut and dried as the mathematical one, but it is much more like it than not. Ethics and social relationships are not random outcomes that could have been otherwise; nor are they purely instinctive behaviours whose only *raison d'être* is to promote reproductive success. Rather, they are thoughtful and informed behaviours which gradually match better and better to the pattern that enables communities of persons to flourish—and that pattern is unrelated to genetics. Our genetic history is not quite as irrelevant here as it is to the truth or otherwise of Pythagoras' theorem, but it is nearly as irrelevant as that.

What this realization does is open up for us the opportunity to inhabit our life as it truly is, and to inhabit it more fully. We can rise to the stature that we in fact possess: the stature of moral beings, with hope and insight, discernment, responsibility and value.

The second main point to be discussed is that biological evolution is highly constrained by many principles of physics, chemistry, sociology and the like and as such it is not direction-less. It is like a process of exploration, and what it discovers in its exploration is an interplay of the random and the structured. The big principles that describe possible kinds of life are not random; some of the details of organisms that express those principles are random. In view of the fact that the process is so patterned and not arbitrary, it is reasonable to hold that it may also be pur-poseful. Scientific study does not drive us to the conclusion that evolution is achieving something purposeful and meaningful; my point here is merely that scientific study is unable to refute the intuition that it is. And indeed it can be put a little more positively than that. Scientific study meshes naturally with other areas of human perception, such as our insights into questions of fairness and justice, value and beauty. This meshing or interlinking is of a particular kind, with the result that the intuition that our lives are part of a meaningful whole is entirely reasonable and indeed a natural conclusion. It is a conclusion which the evidence invites. This is not a proof but it is an invitation.

Thus, the present book aims in part to present reasoning relat-ing to evolutionary biology. I also want to tackle another area of modern science which I think has been misrepresented by experts writing for the general public. This further point (treated first in the chapters that follow) is the physics of the very early universe. I want to point out that physics does not say, and is incapable of saying, that the universe sprang into existence spontaneously from nothing. This is because physics, like all of science, is essen-tially *descriptive*. It *describes* what exists physically. It has the tools of a literary critic, to understand and appraise the structure and dynamics of an existing work, but it does not have the tools to say how the work came to be. If the universe is a story, then the science of physics cannot say how one plot as opposed to another got to be singled out for publication, and it cannot describe the publishing process. I will spell this out more fully in Chapters 5 to 10. On the journey we will meet some fascinating things, such as quantum fields and the transformation of space–time. We will also touch on multiverses and the difference between science and science fiction. And we will briefly contemplate what reason has to say about our ultimate origins.

The common theme of the book is the question of the public presentation of science. This is an important strand in modern culture and we need to get it right, because it contributes to our grasp of questions such as *who are we really?* and *what is our role?* Science does not answer such questions in full, but it provides part of what we need to know. If we get science education right then we all—business people, young people, politicians, religious leaders, parents, educators etc.—will talk to one another in a more well informed way, and then our decisions and values are more likely to be sound. But too much of modern science writing in physics and biology has become a sort of game in which one writer or another starts out from established knowledge and then launches into a flight of fancy, just barely retaining some residual link to research-quality work. The internet and the rapidity of modern publishing often result in such ideas being presented to non-experts as settled knowledge.

I have also felt that sometimes writers have hitched their science writing to a hidden agenda; they are 'telling stories' in the sense of putting across a world-view, and sometimes getting pretty close to 'telling stories' in the sense of misdirection and sleight-of-hand.

But the story of the universe and the story of life on Earth are telling in the other sense of the word: they are striking, and marvellous, and suggestive, and they deserve to be presented as clearly and honestly as we can manage, and told in a full and balanced way. In a particularly noteworthy example of this, I judge that teaching of evolutionary biology will be accomplished better if it is regarded as religiously neutral, rather than trying to hitch it up to atheism or theism, or, conversely, trying to hitch atheism or theism to it.

So here I will present the above-mentioned areas of science, and also the related area of public education and intellectual leadership. I will try to preserve the science, promote the education, and reform the leadership.

All this will lead up to a foray into an issue of plain justice which is signalled by a story about a place called 'Brightland' in Chapter 20, and discussed in the later chapters of the book. Brightland is a place I do not want to live in. Like the Anglo-Saxon poet who wrote *Beowulf*, I feel strongly that leaders should not lay claim to ownership of the common land. The common land here is science and what it shows us about our place in the world. Scientific insight into our physical nature is part of any well-informed and considered overall view of the nature of reality; it is not the private property of naturalism or atheism or humanism or any other ism. So we need to liberate science from this attempt to take it into private ownership, and we need to liberate ourselves from the kind of mindset that has tried to do that.

The subject matter naturally touches on wider aspects of human life, including curiosity about our very existence, and such near-universal impulses as both gratitude and dismay as we grapple with our experiences. I have woven into the text a few reflections on this, mostly with a view to clarifying some basic pieces of religious vocabulary. To echo Charles Darwin, I write in hopes of diminishing the amount of prejudice with which this area is currently beset. And I finish with some pointers to a way forward—how we can recover greater freedom to think, to live and to be, and celebrate both science and human life in all its richness.

CHAPTER 2

WHAT IS A QUANTUM FIELD?

We begin with physics. Our first destination is the physics of the early universe, and in order to get there we will need to consider some physical concepts. I hope to make the argument accessible to a reader with little or no scientific training.

In this present chapter I want to give the non-expert reader a feel for one of the central concepts of modern physics. It is the concept of a 'quantum field'. The word 'quantum' here is serving to indicate the type of mathematics that is needed; the term 'field' is being used to mean something that is extended in space and time, like a stretched-out piece of fabric, or the surface of an ocean, as opposed to something localized at one place like a small, sharp stone or a grain of sand.

Developments in physics for over a century now have been showing us that much—perhaps all—of the physical universe is made out of these quantum fields, or consists of a set of these fields.

Back in the nineteenth century the picture was different. It used to be thought that the universe is made of lots of particles jostling in space, and the regions in between these particles were thought to be empty, unless there is some light passing through. In a dark spot—in a shadow, or far from any star in the vacuum of space,

Liberating Science. Andrew Steane, Oxford University Press. © Andrew Steane (2023).
DOI: 10.1093/oso/9780198878551.003.0002

in the gaps between whatever particles of matter may be floating around—there is nothing there at all, it was thought.

It turns out that this picture is wrong; it is not adequate to capture what has been discovered in physics over a hundred years of steady progress. What we now think is that the situation of matter and space is more like the case of waves on an ocean. The matter in this analogy corresponds to the choppy water where waves toss and crash; the space between the matter is like smooth, calm regions where the water is undisturbed. But these calm places are not empty: the ocean is still there. In a similar way, in between the bits of matter floating in the void of space, the quantum fields are still there.

Wave your hand. You are not moving a lump of something to and fro across a nothing. You are causing a gathering in the fabric of the universe to be worked to and fro like a wrinkle in a carpet or a ripple of excitement running through a crowd.

Come with me and contemplate for a moment a quiet pool of water. We stand by the pool, and if all is calm then the surface of the water is still. Now imagine running your finger or another object through the water for a moment. You will create ripples and vortices. The vortices are little whirlpools and they can be surprisingly long-lived. They are harder for us to see than the ripples, but they are there, and they can have interesting behaviours such as bouncing off one another. The ripples meanwhile spread out, bounce back and pass right through one another.

In this experiment with the pool, the ripples and vortices are 'things' which were not there before we disturbed the water. You might say we have created them. But be careful: by disturbing the water we did not do much in the way of creation; we did not decide the speed with which the ripples move for example, nor the way the vortices interact with one another. Those properties are dictated by the properties of water and the local gravity.

Now let's bring this analogy to bear in considering quantum fields. The quantum fields go by the names of electromagnetic field, quark field, gluon field, lepton field, neutrino field, Higgs field and so on. Each one fills the whole three-dimensional volume of space across the universe, like an immense pool. Together they make up a kind of fabric, one which is much more complicated than the simple ideas about fabrics and fluids that we form from our everyday experience. To understand the quantum fabric, many types of property have to be brought in, called by names such as charge and spin and mass and lepton number. The quantum fabric can be imagined as a multi-dimensional entity, almost as if at each point in space there is a crystal with numerous facets and stresses, but these crystalline fields are the very fabric on which the things of the universe are supported as vibrations or excitations.

When the fabric is not vibrating we do not notice it. But it is still there. To understand why we can assert that—why we do not say that space is mere nothingness in between the bits of matter— notice the properties that the fields have. I used the word 'crystal' just now as way of evoking *precision* and *regularity*. The fields are smoothly spread through space and time, but they have many properties (mass, charge, spin, colour, flavour, lepton number etc.) and these properties are very precise. Now consider one of the particle-collision experiments that are carried out in various laboratories such as the one at CERN, Geneva. In these experiments energy is carefully focused into a small volume of space, and out of that volume there emerge large numbers of tiny particles. But we don't find that just any old particles come out. It is always electrons and groups of quarks and things like that: things with precisely the properties that are dictated by the quantum fields. This is because these particles are not produced from nothing or nowhere, but rather they are excitations of that which is already there: that specific and precise set of quantum fields which has physical existence. The particles are like the ripples and vortices in the pool. You cannot have just any old ripple or vortex, but only the type which the pool can support. Similarly, the quantum fields constrain what kind of matter gets to exist, because that matter *is* none other than an excitation of those fields.

All this will, I hope, enable you to see that to speak of absence of matter as if it equates to absence of physical reality itself is misleading. Indeed, I would say it is misleading enough to amount to a misdirection as complete as the claim that absence of ripples on a pool equates to absence of the pool.

Let's summarize the argument behind the analogy we have been drawing. The inference from particle-collision experiments is that the properties of things like electrons and quarks and photons are not conjured out of nothing during a collision experiment. Rather, the particle collision has caused a disturbance of something that is there already—the quantum fields—and consequently the patterns of mass, electric charge, angular momentum and so on which emerge are those which the quantum fields carry. If we deny that the fields are still there in between such collisions, then either we have a monumental and incredible coincidence to explain—that particles of precisely the same kinds emerge again and again from billions of fresh experiments—or else we have to find some other way of expressing the fact that only some kinds of particle can come to exist physically in the cosmos of which we are a part. In either case there is some sort of physical constraint in place, some way in which the properties of particles are constrained and not merely arbitrary. The operation of that physical constraint, whatever it may be, is an aspect of the physical world. It persists over time and space; it has physical implications. In all these senses of the word, it has physical *existence*; it is not randomness, and it is not mere emptiness or nothingness either.

To bring in another analogy, the particle-collision experiments can be compared to striking the metal frame of a piano with a hammer. One will then hear many sounds, but not just any sound. One will only hear the notes to which the piano has been tuned, along with some noises associated with the frame which fade away more quickly. We can also use this observation the other way around, as experimental evidence. The fact that one only hears specific notes (in the musical example) is *evidence* that something is there making the notes. Similarly, the fact that only specific particles come out of collision experiments is evidence that the 'fabric' is there all along.

So far I have referred to a collection of different fields. Efforts in theoretical physics are often directed towards seeing if these can all be regarded as aspects of a single high-dimensional field. Instead of a collection of different fields all overlapping one another, one may suspect that the fabric of the universe is one multi-dimensional field. By 'multi-dimensional', I mean not just that such a fabric fills the three-dimensional volume of space and extends through time, which is like a fourth dimension, but also that it has properties of its own which, from a mathematical point of view, are like further geometric properties—like directed lines and membranes and volumes, but now extended into further, abstract, mathematical dimensions. By using this notion of further dimensions, and by learning what shapes and topologies and vibrations are possible, one can learn about the properties of mass, charge, spin etc. which are possible in any given configuration. The mathematical techniques are very challenging. It is no wonder that only the brightest mathematical minds can even enter this area of study, called *quantum field theory* or *quantum gravity*. It is wonderful, complex, precise, mind-boggling!

This multi-dimensional field which we call the universe was, long ago, in a configuration different to the one it is in now. Now, in the present, it is configured in the gathered excitations we call stars, planets and galaxies, separated by wide regions of relative smoothness, which we call emptiness or vacuum. Long ago this same fabric was configured very differently. It was in a state of excitation everywhere, a state in which the whole universe was like a hot, dense gas, like the interior of a star. Earlier still, the fabric was configured differently again. It may have begun in an outwardly very calm condition, with no waves, but with a hidden tension, a pent-up energy waiting to be released. It is this configuration which is sometimes called a 'vacuum'. But this is a very special type of vacuum, one that is ready to rip abruptly into a very different state, as it evolves according to the precise laws of its own characteristic motions. I will say more about it after the next chapter.

CHAPTER 3

QUANTUM FLUCTUATION?

In this chapter we will extend our understanding of quantum fields a little, by considering one aspect which is commonly misunderstood. This will be needed in order to understand what is going on when some physicists claim or imply that science can give an account of the origins of the universe—the subject of the next few chapters.

It is often stated, in popular presentations of physics, that the vacuum of space is full of 'quantum fluctuations'. It is said that the fields I described in the previous chapter are always and everywhere in a state of fluctuation. Sometimes it is said that particles and antiparticles are continually 'popping out of the vacuum' and then quickly vanishing again. Sometimes space is said to be a seething mass of such behaviours. Countless books and lectures talk freely about 'jitter', 'bubbling and rippling' etc.

This picture is wrong.

In fact, the quantum fields are completely calm, dormant and not fluctuating at all when they are in their lowest energy state. There is no irregular motion, and all the physical properties are completely static. They are not changing with time, and not varying with spatial location either. The technical name for this is *Poincaré symmetry*.

Liberating Science. Andrew Steane, Oxford University Press. © Andrew Steane (2023).
DOI: 10.1093/oso/9780198878551.003.0003

The reason why the idea of fluctuation has arisen is because it is being used as a way to convey an aspect of the fields whereby some properties are distributed across a range of values, so if one sought to determine precisely the value of such a property, then one would get a range of answers, and this might look like fluctuation. Really it is a case where a still but spread-out thing is being probed by a narrowly directed thing. The situation can be compared to someone standing at the South Pole of planet Earth and trying to use a magnetic compass to find the direction to the North Pole. The compass may wander in all directions, or it may randomly indicate now one direction, now another. But this is not because the direction to the North Pole is fluctuating: it is simply that all directions are equally valid. In a similar way, the quantum fields in their lowest energy state are not fluctuating, but if one chooses to measure certain of their properties, one will get a range of values. Also, if, through a measurement, one forced the fields into some other state, no longer the lowest energy state, then that resulting state would evolve.

The lowest energy state is called the *ground state*. To be clear, I am referring to the joint ground state of the interacting fields. This state is referred to in the mathematical analysis by using the Greek letter omega (Ω) written in between a vertical line and an angle bracket, thus: $|\Omega\rangle$. This should not be confused with another state, indicated by the notation $|0\rangle$. This second state is sometimes called a 'vacuum state' or a 'ground state' but one should be careful: it is not the ground state really, it is another state that is introduced into the analysis for mathematical convenience. The zero state, written $|0\rangle$, is the state which *would* be the one of lowest energy *if* the fields were not interacting with one another. But they are interacting, so $|0\rangle$ is not the true ground state. It is not the state of the fields in the vacuum of space. You can compare this to the case of two balls sitting in a bowl. If the balls do not affect one another, then both will roll to the bottom of the bowl. But if the balls repel one another then their lowest energy location will leave them somewhat apart, with neither of them at the bottom of the bowl. This is an illustration of the fact that $|\Omega\rangle$ is not the same as $|0\rangle$.

To repeat what I have already said, the physical vacuum state $|\Omega\rangle$ does not fluctuate. It is entirely smooth in both space and time. If we ask what are the properties at one place, and then what are the properties at another place, we will get precisely the same answers, so we deduce that in the state $|\Omega\rangle$ the fields are smooth like silk, not rough like sandpaper. They are also smooth over time: the state does not exhibit any change at all from one moment to the next, except that in the mathematical treatment there is a number called a global phase which increments smoothly as time goes on but has no observable physical effect. This complete absence of change over time follows from the fact that this lowest energy state has a well-defined energy, and satisfies the equation called Schrödinger's equation. As experimental evidence of all this smoothness, we can look at the motion of some very sensitive charged particle such as an electron. If we introduce an electron into otherwise empty space, then, once set in motion, the electron moves smoothly, with an exactly constant value of whatever velocity it started out with. If the fields around it were fluctuating then this would not be so. As further evidence, we can notice that some of the electromagnetic waves arriving on Earth have travelled ten billion light years across the universe while deviating by angles of less than a fraction of a degree.[1]

But the term 'quantum fluctuation' is commonly used by physicists in their professional work in this area. So what is going on?

The physicists in their professional work are using terminology in a technical way, one step removed from its meaning in everyday life. Unfortunately, this particular piece of terminology is not very well-chosen because it leads to the sort of misunderstanding that I have been describing, and we will see later that this issue is particularly relevant in the early universe. Arguably, it might be better to drop the term 'fluctuation' altogether, and employ something else such as 'superposition' or 'spread'. But I doubt that will happen; physicists are human beings and they have fallen in love with the phrase 'quantum fluctuation' and its vaguely exotic connotations.

For the avoidance of all doubt, I am not in this chapter denying the existence of what are called 'quantum fluctuations'; I am simply pointing out that they are not fluctuating. It's like someone who wanted to point out that the planet Venus is a planet not a star. If someone says 'the Morning Star is not a star', they are not denying the *existence* of Venus, they are just saying it is not a huge ball of fusion plasma. Similarly, when I say the quantum superposition associated with fields and other things is not fluctuating (in a state of given energy) then I am not denying the existence of the superposition, I am simply denying that there is anything changing from one moment to the next.

A related term, also somewhat unfortunate, is the phrase 'zero-point motion'. This term is a hangover from the early days of quantum theory, when physicists were trying to interpret what

---

[1] For experts: this is a statement about angular blurring in the cosmic microwave background.

the equations were telling them. When one enquires closely into the behaviour of any physical thing, be it a ball, a sound wave, a magnetic field or whatever, one finds that its properties have the spread-out nature that we refer to with the phrase 'Heisenberg uncertainty principle'. So, for example, when a pendulum hangs still, it is not quite true to say that it hangs in the vertical direction and that is all. Rather, it is aligned on average vertically, but its situation includes a spread-out distribution including angles close to the vertical. One way to imagine this is to compare it to the case of a pendulum swinging to and fro over a small range of angles, and hence having a residual motion called zero-point motion. However, this picture is just an aid to the imagination. It should not be trusted too completely. It is like a parable or a myth: it contains some truth, but that truth does not consist in taking the story literally. In the example of zero-point motion, there is not in fact any sense in which the system is evolving over time; there is no motion in that sense—no change, no development.

For an analogy, suppose an amateur detectorist goes out on an expedition with his daughter, and, pointing his new long-range, wide-angle metal detector towards the ground, he finds that the device detects the presence of metal, but the direction indicator is surprising. First it points left, then right, then straight down, then left and right, never settling. 'Strange', he says to his daughter, 'the metal seems to be randomly moving to and fro under the ground'. 'Err, dad', she replies, 'couldn't it just be a long pipe?'

The main message of this chapter is that the quantum fields in their lowest energy state, called ground state, do not have dynamically fluctuating behaviour. But the reader may have had the chance to see one of the videos that some teams of theoretical physicists have used to illustrate the physics of this quantum state, and those videos show much bubbling and rippling—lots of fluctuation. Are they wrong, then? They are not totally wrong, but they are not telling the whole story, and neither have I up to this point in the chapter. The whole story includes the following fact: in quantum physics, two motions can add up to no motion. But if this is so, then, equally, no motion can be regarded as a combination of two motions!

Consider the situation of two lamps shining light at one another. Say lamp A is on the left directing its beam to the right, and lamp B is on the right directing its beam to the left. Light is passing from A to B, and light is passing from B to A. So, overall, is the light going anywhere? Is it moving? There is a sense in which it is both moving and not moving. If we measure properties such as the net energy flow at any location in the combined light beam, we get zero. But this zero is not because there is no energy at all, and no movement at all—it is a result of an equal amount of movement from left to right as from right to left.

The situation with the quantum vacuum is similar. There is no net motion, but there is a rich internal structure, such that the total can be seen as the result of the combination of many complicated motions. I will develop this thought further in the next chapter. These complicated motions 'hiding' in the static whole might be regarded as fluctuations, so one might want to say that the state is fluctuating in some sense after all. In the end one cannot insist on which use of words is a fairer attempt to convey the physics. But some aspects of the physics are clear and sharp, and one such aspect is that quantum fields in their ground state cannot suddenly start to do new things such as start to vibrate or start to generate particles. If the term 'quantum fluctuation' suggests to non-experts that there could be such a spontaneous movement, a developing evolution as a result of a random fluctuation, then the non-expert is mistaken and perhaps the terminology invited the mistake. Such a development in the fields in their ground state does not and cannot happen. The state is dormant in this sense.

In Chapter 5 we will consider a case where the state is like the lowest energy state in some respects, but not in all respects. In that case real dynamic evolution can occur, as we will see. But before getting to that, we shall have an interlude, where I try to unpack some of the popular ideas about quantum fields.

CHAPTER 4

THE VACUUM AS A DYNAMICAL SYSTEM

In the previous chapter I introduced the symbol $|\Omega\rangle$ to refer to the situation for the collection of quantum fields in their lowest energy state. In this state there are no particles, and furthermore, no particle–antiparticle pairs ever suddenly appear and then quickly disappear again. So why have many people asserted that they do? It is because they are trying to convey to the layperson an impression of the marvellously complex and structured nature of the empty vacuum of space. What is empty in one sense (there are no particles or excitations) is full in another (the fields are all there, pulling and pushing on one another). But why is this often described, in popular presentations, as a picture of particles and anti-particles popping into and out of existence? This is because a certain calculational method could be seen that way, if you squint a bit. But I think most experts would agree that this way of putting it relies on an extreme looseness of language, where that which is not a particle is called a particle, and that which does not fluctuate is called fluctuation. But if we allow ourselves to speak as loosely as that, then we will not be undertaking science communication; we will be doing something else: passing off as science that which is not faithful to science. The calculational method behind these ideas is perfectly correct and indeed precise

Liberating Science. Andrew Steane, Oxford University Press. © Andrew Steane (2023).
DOI: 10.1093/oso/9780198878551.003.0004

and useful. The question is, how can we put it into words fairly and squarely? And what does it tell us about the nature of the vacuum?

There are two items of technical terminology which one has to get straight. These are the terms 'bare particle' and 'virtual particle'. And don't forget the underlying theme that has to be kept in mind: what we call a 'particle' is an excitation of one or more fields.

Let's begin with 'bare particle'. A 'bare' particle is a mathematical idea. If we take the mathematical description of a real particle such as an electron, and then cross out of the equations certain parts, then what remains is a mathematical concept which can be said to be much like an electron, but one with no connection to the electromagnetic field. It has electric charge but it doesn't make an electric field and it doesn't interact with light or radio waves etc. This is an abstraction that has been arrived at by eliminating some features from a mathematical description, as I have said. It is called a 'bare' electron.

I will now develop an analogy which I think does a fair job of capturing the difference between a bare electron and an actual electron. I will call the latter a 'full electron'. A full electron is the very thing which in all the rest of science is called simply 'an electron'. In quantum field theory this full electron is called a 'dressed' electron to distinguish it from a bare electron.

For purposes of building an analogy, let's consider an elastic membrane such as a trampoline. This is an elastic membrane on which you can bounce, and which can sustain ripples and other oscillations. We can imagine a trampoline being made out of two types of elastic cord, woven together in the pattern called 'warp and weft' in the manufacture of ordinary fabric from thread. One set of cords, the 'warp', runs north–south, say, and the other set, the 'weft', runs east–west. Assume also that the two types of cord are different: say one is thicker than the other. This trampoline is, in the analogy I am developing, a picture for the combination of two fields: the electron field and the electromagnetic field. Excitations of the first field are called electrons and positrons; excitations of the second field are called photons. So far so good. But what you can see immediately is that when you excite the warp (the electron field), you also excite the weft (the photon field), because they are woven together. If you move the cords running north–south, then those running east–west move as well. In a similar way, an ordinary charged particle such as an electron is really an excitation or disturbance of *both fields together*. So what we called the 'electron field' is not, on its own, the whole story about electrons. And what we called the 'photon field' is not, on its own, the whole story about photons. But for purposes of calculation we could imagine that the warp and the weft were separated and not woven together. In that case you could have a disturbance of the warp which did not disturb the weft. This is the idea called 'bare particle'.

The change from disconnected warp and weft to woven warp and weft is like the change from $|0\rangle$ (the zero state) to $|\Omega\rangle$ (the actual state of the fields in empty space). It is found, by calculation, that the interaction part, where the fields are woven together as it were, introduces a change throughout the fabric, like an added tension, so $|\Omega\rangle$ is very different from $|0\rangle$. In some respects it is infinitely different.

The difference between $|\Omega\rangle$ and $|0\rangle$ sometimes catches out experts, because the state $|\Omega\rangle$ can be written as a combination of $|0\rangle$ and other states: states with more energy than $|0\rangle$. So this suggests that there are, in $|\Omega\rangle$, various energetic excitations all combined together. Those energetic excitations could in principle be detected, the argument goes, and this amounts to detecting particles. This argument is wrong because such excitations are bare electrons or bare positrons or bare photons. They are not actual electrons or positrons or photons and they cannot, even in principle, be detected. They are a sort of logical abstraction, like thinking about how a planet would be if it had no gravitational effect. But a real planet does have, and cannot avoid having, its gravitational effect. Similarly, a real charged particle does have, and cannot avoid having, its electrical effects. So the 'bare' electrons are not electrons. A bare electron is not the thing which twirls around in atoms and moves along electrical wires and makes clicks in detectors. The thing which twirls around in atoms and moves along electrical wires and makes clicks in detectors is the full electron, the combined excitation of all the relevant interacting fields, and that thing is not present when the state is $|\Omega\rangle$.

The other ingredient we need to discuss is the idea called 'virtual particle'. The term 'virtual particle' refers to a useful way to organize some calculations in quantum field theory. One does not need to use that method, but it can be very helpful. It can help both in calculating and in gaining physical intuition. But how to express this concept to the non-expert? The calculation is a way to set out the interaction of the fields; the weaving-together of warp and weft in my analogy. It can be useful to write the calculation as a sum of terms, like the sum

$$1 + \frac{1}{2} + \frac{1}{4} + \frac{1}{8} + \frac{1}{16} + \frac{1}{32} + \frac{1}{64} + \dots$$

(one plus a half plus a quarter plus an eighth, etc.), only in the field calculation the terms in the sum are a good deal more complicated than this. To organize the sum we draw a set of diagrams. Each diagram shows a network. Each such network tells us how to write one term in the sum. The nodes and lines in the network correspond to mathematical factors which when multiplied together give a function which depends on the locations of the nodes. This function then has to be added up (integrated) for all possible locations of the nodes, and thus one obtains the term we are calculating. And this is just one term in an infinite set of terms!

In these calculations, each line in a network corresponds to a disturbance in the fields with a specific amount of energy, momentum, electric charge and some other properties. It is these lines which are often called 'virtual particles'. One thing to note, then, is that a virtual particle is like a link in a network: it cannot exist independently of the network. It has *betweenness* not *thingness*. Another important fact is that this is no ordinary network, because each node does not sit at any particular place or time. Instead, each node represents a contribution which is smoothly spread over all places and all times! I hope it is becoming clear that these networks are first and foremost mathematical abstractions. They help us to organize calculations. But they also have a rather striking resemblance to networks of real particles whizzing about and bumping into one another. So what should we say? When fields interact in the vacuum, are there particles whizzing about or aren't there? The least misleading answer is probably 'that depends on what you mean by a particle'. And if what you mean by a particle is the thing that moves along in electrical wires and twirls around in atoms and makes detectors click then the answer is 'no, there aren't any of those in the vacuum'.

I already mentioned one important difference between real particles and virtual particles: the internal lines in the abstract network (i.e. the virtual particles) are precisely that: internal. They have no impact whatsoever on anything outside the network. Their only contribution, even in principle, is to help express how the lines representing real particles coming into any given network relate to the lines going out. Another difference concerns energy and momentum. The field disturbance corresponding to any internal line has a combination of energy and momentum that no real electron or photon could ever have. This difference alone would be enough to show that the contribution represented by one of these lines is only called a 'particle' by a stretching of language. One consequence is that these disturbances propagate differently from real particles: they fade away exponentially as they go. When one combines this with the other features, one has to question whether the word 'particle' is really helping at all. As theoretical physicist Matt Strassler of CERN has put it in his blog, 'Of Particular Significance',[1]

> The best way to approach this concept, I believe, is to forget you ever saw the word 'particle' in the term. A virtual particle is not a particle at all. It refers precisely to a disturbance in a field that is not a particle. A particle is a nice, regular ripple in a field, one that can travel smoothly and effortlessly through space, like a clear tone of a bell moving through the air. A 'virtual particle', generally, is a disturbance in a field that will never be found on its own, but instead is something that is caused by the presence of other particles, often of other fields.

---

[1] See https://profmattstrassler.com/articles-and-posts/particle-physics-basics/virtual-particles-what-are-they/.

Having said all this, I am not about to insist that we drop the term 'particle' in 'virtual particle'. I don't mind about that. And I welcome the fact that physicists have introduced the general public to the notion of virtual particles. For the present chapters I don't mind whether one adopts the method of virtual particles or not. What I am mainly wishing to question is the use of the term 'fluctuation'. What I want to underline is that virtual particles are not fluctuations in the ordinary sense of the word. They are contributions to a whole and their contribution is certain, not probabilistic. The calculation is comparable to the way many small tensions between fibres add up to make the total tension in the fabric of a trampoline.

For the ground state, $|\Omega\rangle$, the overall result, the result of the entire network of interactions, is smooth and dormant, no matter how it is calculated. Once we have allowed the warp and weft (in the trampoline analogy) to engage with one another and pull the fabric together, the result, in its lowest energy state, is stationary. It is a stationary state.

For completeness, I should also mention that there is one last gasp of an idea whereby someone might want to maintain that particles can pop into existence in the vacuum. This involves the fact that, in a finite length of time, frequencies are never precise, and therefore energies are not precise either. Suppose we consider an experiment where one introduces a detector which is switched on for a short period of time. Such a situation cannot have a precisely defined energy, so such a detector is not guaranteed to have its lowest energy; it might have a bit of extra energy provided by whatever process set it up in the first place. In this case the detector can itself provide whatever energy is required to form photons or particle–antiparticle pairs or other things. In that case the photons or the pairs were not spontaneously appearing out of the vacuum, they were formed from the disturbance introduced by the detector.

A detector for electromagnetic waves—a photo-detector for example—must be allowed to settle for some time before we can consider that its output is trustworthy. If we were trying to detect photons at microwave frequencies then the wavelength is a few centimetres and the frequency is around ten gigahertz. The associated wave completes an oscillation in one tenth of a nanosecond. This means the detector cannot be deemed reliable until many tenths of nanoseconds have passed after switching it on. Before that the detector might signal that it has one photon's worth of energy more than its ground state, but a competent experimentalist will know that this energy in the detector was part of the way it was prepared. The energy did not come from the surrounding vacuum, it came from the process of setting up the detector. Once enough time has passed for the detector to settle down to a well-defined energy, thereafter it reports no further incoming energy if it is surrounded by vacuum: no clicks, no photons, nothing.

I included the previous paragraph in order to forestall anyone who says that energy–time uncertainty can lead to particle formation in the vacuum. The reply is that this route is available, but only if you do not in fact have a vacuum! (Or you have one only for a short time.)

If someone shows you a picture of the vacuum, and in the picture there is anything like little dots or little motions, then you should understand that the picture has adopted a good deal of artistic licence. It is like the pointillist technique adopted by artists such as Georges Seurat and Paul Signac. In this technique a continuous surface such as a human face is shown by a collection of dots. This allows the artist to achieve certain effects, not the least of which is to remind us, subconsciously, that we are looking at an impression, not the original person or scene, and the vision being offered to us is all about the continuous lines and shades which the dots suggest. In a pointillist work the dots are precisely *not* the point! They are servants, not the master. They serve to convey the whole without pretending to be the whole.

We are not misled by pointillist artworks because we already know the whole to be smooth. In a similar way, don't let the little dots or motions in someone's picture of a quantum vacuum mislead you. They are trying to show you something which is smooth really, and in which all motion in any given direction at any given place is balanced by an equal and opposite motion at that very same location in space.

CHAPTER 5

THE VERY EARLY UNIVERSE

Modern physics tackles the question of the large-scale structure and history of the universe using two main lines of evidence: astronomical observations, and theoretical treatments which enable one to construct a model which fits those observations. The theoretical treatments draw on quantum field theory, and on the theory of gravity—Einstein's general relativity. These theoretical frameworks are in turn supported by many observations of particle interactions here on Earth, and by precise measurement of various gravitational phenomena such as planetary orbits and gravitational waves.

The big picture of the universe is that it is vast, and ancient, and the distances between the galaxies are getting larger as time goes on. This fact is known as the cosmic expansion, and it is consistent with the model called the *Big Bang*, in which at some very early time the universe was in an extremely hot, extremely dense state, from which it grew explosively and cooled as it grew. These statements are very well supported by large amounts of evidence, which I shall not elaborate on here. My aim in this chapter is to present what is and is not known about the situation at the very earliest moments, where the evidence is not so abundant or clear. My aim is to give the flavour of the types of model which are suggested by modern physics, and also to inform the

Liberating Science. Andrew Steane, Oxford University Press. © Andrew Steane (2023).
DOI: 10.1093/oso/9780198878551.003.0005

reader concerning the degree to which our knowledge here is provisional and uncertain.

One can be most confident of our understanding of physics in areas where it has been studied and tested empirically. For example, experiments in the Large Hadron Collider at CERN, Geneva can reproduce conditions of temperature and density that are comparable to those reckoned to have held in the universe at large when all the matter distributed throughout the universe had that same temperature and density. This temperature can be roughly estimated by taking the collision energies achieved in the Large Hadron Collider (about 10 tera-electron-volts or one micro-joule) and dividing by a standard conversion factor called Boltzmann's constant. One finds a temperature of 10^{17} kelvin, which is about

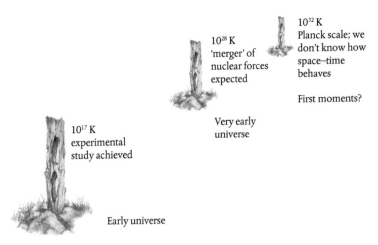

10^{28} K
'merger' of
nuclear forces
expected

10^{32} K
Planck scale; we
don't know how
space–time
behaves

First moments?

Very early
universe

10^{17} K
experimental
study achieved

Early universe

Figure 5.1. A sequence of boundary posts going back to the earliest moments of the universe. The conditions attain higher and higher densities and temperatures, and our scientific understanding becomes more and more tentative.
Source: www.joshuanava.biz.

six billion times the temperature at the core of the Sun. (The *kelvin* is a unit of temperature; at high temperature you can read it as 'degrees Celsius' if you like. 10^{17} kelvin is one hundred million billion degrees Celsius.) This estimate gives a 'boundary-post' as we explore back towards the very early universe. As we explore back to earlier times, we are entering territory that is more and more unknown. The boundary post at 10^{17} kelvin marks the end of human knowledge that has been subjected to detailed empirical checks.

The tools of quantum field theory enable us to extrapolate back to earlier times when the universe was hotter. We can extrapolate with reasonable confidence back to the time when the temperature was that at which the various non-gravitational forces (strong, weak, electromagnetic) are expected to merge into one another, and show themselves to be different aspects of one idea. This 'merging' has already been found to happen for electricity and magnetism and for the electromagnetic and weak nuclear forces, and our understanding of quantum field theory is such that it is confidently expected that it will also occur, at sufficiently high temperature, for the strong nuclear force. However, it has not yet been possible to formulate a theory that expresses such a merger while retaining overall consistency with observations (at the lower energies we can access). Actually, that statement does not capture the whole situation. What happens with some treatments, string theory being the prime example, is not that they express too little but that they express too much: the mathematical framework may possibly capture correctly what goes on in the universe, but we do not yet know how to pin it down, mathematically speaking, to single out the observed universe from all the other possibilities that the mathematics can describe.

The energy at which the merger of the various nuclear forces is expected to happen corresponds to temperatures around 10^{28} kelvin. This temperature is a kind of first stage-post out in *terra incognita*. It sits out there well beyond the boundary of what we precisely know, and at the energy scale where substantially different behaviour is likely. The attempt to reason about processes at this temperature involves an extrapolation by a huge factor from the boundary-post at 10^{17} kelvin provided by the Large Hadron Collider. This extrapolation is by a factor of 10^{28} divided by 10^{17} which is 10^{11}, that is, one hundred billion. In the history of physics, it has often[1] been the case that new and unexpected things were discovered as experiments gained access to higher energies and temperatures, so one should be wary of making big extrapolations based on current knowledge: they may be missing out things that have not yet been dreamed of by us. This should be kept in mind when reading discussions of the very early universe.

The 'merger' theories are called 'grand unified theories', or GUTs for short. Everything we say about the universe before it cooled to the GUT temperature has to be regarded as tentative, and very likely will miss significant components. But what we can attempt already is to formulate reasonable qualitative statements about what sorts of process may have gone on.

The next post, situated way out in the unmeasured wilderness, is at 10^{32} kelvin. This is the temperature where energies are so high that, owing to the curvature effects predicted by general relativity,

---

[1] The word *often* could be replaced by *always* here and it would still be a fair summary of the history! One of the current unresolved puzzles concerns something called *supersymmetry* and the *hierarchy problem*.

space–time itself has structure at a distance scale that cannot be treated by classical physics. This means that some sort of quantum theory of the whole of space–time and the matter in it needs to be invoked. The theories known as *loop quantum gravity* and *string theory* (or M-theory) are the two most well-known attempts to grapple with this. This earliest era in the unfolding of the universe is called the *Planck era*. I call it 'earliest', but strictly it is an era in which the notion of time itself becomes changed. In the Planck era it is not clear whether or not terms such as 'before' and 'after' have meaning. And we do not know, and we have no way of knowing, whether this era was itself the outcome of some physical process or other.

Another aspect of the universe which is unknown to us, and which may remain permanently unknown, is whether or not the universe is infinitely extended in space. When some physicists describe the universe as if they knew it to be infinite, they are making statements they cannot know to be true. Modern physics asserts that a universe of finite spatial extent is a perfectly coherent state of affairs which need not imply the presence of an edge where space abruptly stops. This is because general relativity allows various possibilities for the large-scale topology (i.e. the way the universe connects to itself), and this is true whether the universe is curved in either a positive, zero or negative sense. In short, if we make the guess that the universe is finite, we are not thereby guessing the three-dimensional equivalent of a flat disc which stops at an abrupt edge. Rather, we would be guessing that the universe is the three-dimensional equivalent of the surface of a sphere or a torus or something like that; something which folds back on itself and therefore never needs to come to an edge.

Transitions in the very early universe

Let us now consider what may have taken place in the Planck era, and in the very early universe immediately after the Planck era. As I have said, we can try to formulate reasonable qualitative statements about what sorts of process may have gone on during and after the Planck era. We do this by proposing a model of the quantum fields as described in the previous two chapters, but now our model has to account for gravitation as well.

The mathematics that describes the dynamics of these fields, with gravitation at work, is itself rich and beautiful. One of the concepts that arises in this area has been given the technical name *false vacuum*. The notion of false vacuum has been and continues to be a very interesting development in field theory. One finds that the equations that describe fields, whether in a classical or a quantum treatment, allow that the vacuum of space can be found in a state comparable to that of water vapour cooled below the dew point. Such a state is metastable, meaning that it can persist for a while, but will eventually evolve into a different state. In the case of water, one has a vapour, but at any moment droplets of water can begin to form, and once they do they can grow until the whole system has turned into a liquid. In the context of the vacuum of space, the equivalent process can lead to a transformation between a state with no particles and therefore empty of matter, to a situation where the least energetic state of the fields has moved to lower energy but the total energy of the fields has not. It follows that the fields are now in a condition of excitation, and this corresponds to the presence of large amounts of light and matter.

This is a physical process where the equivalence of mass and energy has allowed energy to be concentrated into the mass of large numbers of matter particles (real ones, not virtual ones), and this energy was in another form at the outset. The energy was there, but one might describe it as hidden. In the initial state (the one called false vacuum) the energy remains in the *skein* or fabric, but in the final state the fabric has found a new configuration, somewhat as if an initial tension has been released, such that stored elastic energy has been converted into vibrations. These vibrations form the particles of matter. This is the type of process that is being described when some physicists talk excitedly of 'something coming from nothing' or 'the creation of the universe'. Actually what has happened is that the universe (the set of quantum fields) was there all along, but something very interesting has happened: it has moved from one configuration to another, and the first configuration does not seem to be much like a universe in the ordinary sense of the word, whereas the second configuration does. It is a bit like having at first a dark ocean, every element of which contains precise structure and connects to adjacent elements through precisely prescribed links, but this structure is hidden until it makes its presence felt through its influence on the dynamics. Then a dynamic morphing happens, one that can begin anywhere in this ocean, as gravitation combines with all the other forces of nature to move energy from one form to another, and form both light and matter out of the fields that support them. One might say that the process 'creates' light and matter, but note that the word 'create' here is standing in for an ordinary case of cause and effect, like the way a loudspeaker 'creates' sound waves in air. The loudspeaker is itself part of the physical universe, and is operating in accordance with specific physical laws, and the sound that results is a reconfiguration of pre-existing material (the air). In a similar way, the gravitating quantum fields pull and push on one another and 'create' matter by causing excitation of the pre-existing extended fabric provided by those same fields. It is part of their physical nature to do that, if initially they were in a false vacuum state. The matter is itself an aspect of the fields when they are in the new configuration. The process is a process of transformation.

When a chameleon changes the colour of its skin, the chameleon is there throughout. Likewise, when matter particles are formed, the quantum fields are there throughout. They are merely changing their appearance.

After such a process, the physical system (i.e. what was the vacuum of space) is now in a highly excited state (i.e. full of particles of matter and light). It might appear as if these particles have come from nowhere, but really they are excitations of the fields which were there all along, and this process of particle formation is part and parcel of the nature of these fields, of the dynamics they display when the initial conditions are very special.

Another analogy for what happens throughout the vacuum of space in this process is the release of water from a dam. Suppose there is a large reservoir full of water, and the surface of the water is calm and still. This corresponds to the false vacuum state. The question arises, is there a way for these still waters to be disturbed? With no wind blowing, it might appear that the answer is no. But then someone notices that the reservoir is not really in its lowest possible energy level. It is being held back by a dam. If a gate is opened in the base of the dam, then the water will rush out with large amounts of kinetic energy, filling up the region below with crashing, jostling waves. In this analogy the 'dam' is present throughout space initially, and then the transition described above 'breaks through' it. The main purpose of this analogy is to illustrate where the energy required to make the matter waves comes from.

To summarize, some such transition in the configuration of the quantum fields is what modern physics suggests for the processes at work in the Planck era or soon after it. After some such process the universe acquired the very hot, very dense state from which it subsequently expanded and cooled. That expansion and cooling can be described by more well-established descriptive frameworks, especially general relativity.

CHAPTER 6

NOTHING COMES
OF NOTHING

In this chapter I will discuss the following simple statement:

> Modern quantum physics and cosmology do not explain or otherwise account for the fact that there exists a physical world (called by us 'the universe').

Most philosophers would, I think, find this statement uncontroversial and even obvious. However, sufficiently many contemporary physicists appear to be confused about it, or to write misleadingly for the general public, that the question merits a fresh comment.

In the ancient Greek epic poem known as the *Odyssey* and attributed to Homer, the protagonist Odysseus and his band of shipmates are captured by a cyclops—a giant who holds them in his cave and prepares to devour them at his leisure. Odysseus offers the cyclops wine, which he likes, and the cyclops enquires who has given him the wine. What is his name? Odysseus replies, 'Nobody'. Later, after the shipmates have blinded the cyclops and he cries out for help, his fellow giants ask who is tormenting him. He replies: 'Nobody! Nobody is doing it.' So they ignore his pleas and Odysseus gets away.

Liberating Science. Andrew Steane, Oxford University Press. © Andrew Steane (2023).
DOI: 10.1093/oso/9780198878551.003.0006

This little play on words makes the story fun and memorable. We, hearing it, enjoy the cleverness of Odysseus getting one over on a bunch of ignorant giants. We relish the double meaning of the word 'Nobody' in the story. Of course, we don't make the mistake of muddling up the meanings as the giants do. We are clever like Odysseus.

Now it seems to me that this sort of wordplay is going on in the modern practice of using the word 'nothing' to refer to one of the states of the collection of quantum fields that extend throughout space and time. Here the term 'field' is being used in the technical sense which I have outlined in the previous chapters. It refers to an extended physical thing which is somewhat like an ocean or a pool or a skein. The collection of these 'fields' together makes up the physical fabric of the universe, according to a well-developed understanding of physics that has been acquired by the community of scientific researchers over a long period. Naming a huge, complex physical system (i.e. the quantum fields, or whatever description underlies them, be it string theory or some other) with the term 'nothing' is understandable as a sort of in-joke for those who understand field theory, but employing it in wider writing directed towards the general public is misleading. It just spreads confusion, and this is not a fair thing to do, because our fellow citizens are not a collection of nasty giants who would as soon eat us as look at us. It is not just poor practice but improper practice to answer, to someone asking how the universe came to be, 'Nothing! Nothing did it'! as if that were anything but a sheer nonsense statement, or else a trick like the trick of Odysseus—the use of the word 'Nothing' as a proper name for that which is not nothing.

In the previous chapter I described the type of transitions that quantum field theory suggests can take place in the basic structure of the universe. The fields can change, like dew drops condensing, and like a dam being opened with a great release of energy. After such a process, where there were previously no excitations—no waves, no particles—now there are many. The still, calm initial state does not last; it is unstable, and quickly moves to the energetic, hot subsequent state. It is this kind of process which was pithily summarized by Frank Wilczek's remark, '"Nothing" is unstable'. Note the capital N. Seen as a remark on the physics of quantum fields for those familiar with the in-joke, this is a fair remark. Considered as a summary for the non-expert public, however, it fails. It sends a misleading message, because people are understandably interested in questions of origins and this language seems to suggest that a physical theory such as quantum mechanics can somehow account for existence itself. This it certainly cannot do, because even to make any statements at all, any scientific theory has first to have some physically existing stuff to make statements about. It is like the instruction 'first catch your hare' in the old recipe book. The quantum recipe can tell us what the hare—the quantum fields—will do. It cannot conjure those fields into existence.

The formal name for this distinction is *metaphysics*. *Metaphysics* is the name for the branch of knowledge which deals in issues such as physical existence itself, and the difference between possibility and actuality, and what constitutes explanation in this area. When we look at a bottle of milk, and ask 'where does it come from?' then someone might answer 'from the supermarket' or 'from a cow'. But these answers do not satisfy all our wonder. An artist gazing at the milk, and drawing a still-life picture, experiences a sense of wonder at its sheer fact of existing. Could it not have existed? Why is it here? Where is it 'coming from' right now in the present moment? Scientific reasoning can connect the bottle to things that preceded it and things that will follow after and around it. The scientific method can uncover the patterns in the tapestry of the universe. But tracing the links *within* those patterns cannot say how such a tapestry comes to be there, nor why, nor that there is no why (if there is not). Those are metaphysical questions.

Returning now to the commentary offered by modern physics, another misleading practice is to say that the universe originated in a 'quantum fluctuation'. This is like saying that a hurricane is caused by a butterfly. Such a statement forgets the 'small' matter of a huge atmosphere with temperature differentials, Coriolis force, Bernoulli equation, gravity, wind and Sun and land and ocean. If the frequency and violence of hurricanes are increasing, it is not owing to a surge in the butterfly population! The reason there are hurricanes is not the flapping wings of insects but because all these powerful influences are at play in Earth's atmosphere, and the weather system is unstable. In a similar way, if the universe starts out in an unstable state, then the cause of the change that follows is to be found in the forces at work in the quantum fields. When one expert in this area of physics speaks to another, they may use the term 'quantum fluctuation' to refer to the combination of trajectories that the quantum fields can follow, or to the spread-out character of quantum states which prevents the fields from remaining permanently balanced in an unstable condition. But to say to a member of the general public that the change is owing to a 'quantum fluctuation' is to mislead them, because the words mean something else to a member of the general public: they mean something like a tiny random movement. This fails to convey the fact that the process is not random at all but determined by the interactions, and it fails to convey the wonderful complex structures that are evolving with perfect precision from one configuration to the next. To say that the early development of the universe is owing to a quantum fluctuation is like saying that a performance of Bruch's Violin Concerto No. 1 is owing to the fact that a violinist had a tremor in his arm.

A further misdirection occurs when scientists imply that there is a fluctuation from nothing to something, from non-existence to existence. Whether or not people intend to mean this, it is what they are often taken to mean. But quantum fluctuations are fluctuations *of* something. They are behaviours of quantum fields that are already in existence. So whereas it is correct to say that fields that exist can fluctuate, it does not follow that there can be a fluctuation between non-existence and existence of the fields themselves. It is strictly illogical to take language which applies to the kinetics and dynamics of physical systems, and transfer it, unchanged, to metaphysics, and expect to get sense and meaning without further intellectual work. One might get sense, or one might get nonsense: it all depends on what meanings are given to the terms when they are employed in the new setting.

When physicists speak of a fluctuation at an early stage in the evolution of the universe, from which a great change came about, what they are saying is that the fabric of the universe (the quantum fields) went through a transition from one configuration to another, as I explained in the previous chapter. This is still a very interesting insight. But it is not any sort of creation of something from nothing, and nor is the fluctuation itself the chief cause of what followed after it. The chief cause is the fields themselves. It is

NOTHING COMES OF NOTHING

their innate nature which underpins the sort of universe we have got. Just as the nature of hurricanes has got very little to do with the nature of butterflies, so, equally, a good understanding of the current state of the universe is not to be had by merely saying, 'oh well, it is just the result of a random fluctuation'.

Suppose an artist were to set up a public artwork which consisted of a huge steel pole standing upright and balanced on a small penny piece, with no other support. It would not be long before a gust of wind caused the pole to topple. Suppose it toppled with catastrophic results: it smashed through the roof of the town hall and narrowly missed a passing pedestrian. Now picture the scene in the courthouse where lawyers are arguing whether anyone can be held responsible for this disaster. Suppose a defence attorney argued, 'It is no-one's fault: It was just a random fluctuation! A gust of wind did it'! Such an argument would, I think, be found to be largely missing the point.

This large steel pole is an illustration of the situation called *unstable equilibrium*. It is the situation where something is unmoving at first, but is liable at any moment to set off at gathering pace towards some other state of affairs. It is the situation of the quantum fields of the universe if, in the Planck era or soon after it, they were in a false vacuum state.

CHAPTER 7

RUBBLE AND RANDOMNESS

Another misleading or illogical practice is to assert that the early state of the fields which make up the universe was itself a state of randomness. You sometimes see or hear physicists describing the early universe as a seething mass of particles and antiparticles randomly popping into existence and disappearing again. I witnessed such an image in a lecture from a leading astrophysicist at Oxford University, for example. The image has caught on: one celebrated writer declares the fabric of space to be, in its earliest moments, 'a high-energy chaotic quantum froth',[1] which it certainty was not. But I don't blame that writer (Randall Munroe). He is conveying what many physicists have said. It is almost entirely misleading. In fact, the fabric of the universe—the quantum fields—is intricate and precise. Far from being random—a 'froth'—it is, and always has been, very highly ordered.

There is a famous composite image/diagram which has been present for many years on the Wikipedia page describing Big Bang cosmology, and has been used in many scientific presentations (Figure 7.1). It was prepared by the NASA/WMAP science team in order to illustrate the evolution of the cosmos, and it is a fine

---

[1] Randall Munroe, *How To*, John Murray 2019.

Liberating Science. Andrew Steane, Oxford University Press. © Andrew Steane (2023).
DOI: 10.1093/oso/9780198878551.003.0007

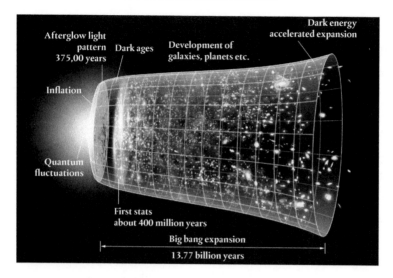

Figure 7.1. A famous and widely used composite image/diagram illustrating the development of the universe as a whole.
Source: NASA/WMAP science team; Wikimedia Commons.

image, one of the finest on Wikipedia. It mostly does an excellent educational job, gathering information about time periods and major events and giving a visual impression of the developing structures in the universe. But, referring to the earliest moments of the universe, the label on the diagram reads 'Quantum fluctuations'. This label is there because the diagram was developed as a teaching aid in the context of explaining the significance of small, random variations in certain microwaves arriving at Earth, which I will describe in a moment. But unfortunately this label becomes misleading when the diagram is used, as it commonly is, as an introduction to Big Bang cosmology. For the early universe was just about the smoothest place you can imagine! What is being labelled 'quantum fluctuations' is here a case of *hyper-precise order*. This orderly motion has the nature of quantum structure, and therefore it includes quantum harmony, sometimes called quantum fluctuations. Here I call it 'harmony' because it is a case of wave-like motion involving many contributing motions at the same time, like notes in a musical chord, and it is a very accurate and specific chord, not a random one. This chord (the technical name is *superposition*) is precisely defined in size and character, and it is part of the physical nature of the intricate set of quantum fields, or single multi-dimensional field, which forms the universe. This is a field or gathering of fields which in its earliest moments *especially* was very highly ordered and in a precisely defined state. (The reason for the word *especially* in the previous sentence will emerge in a moment.)

The label on the cosmology diagram is like preparing a map of the Indian city of Agra, and next to some rough stone providing a label marked 'pile of rubble', all the while failing to notice the Taj Mahal.

I expect the reason why the NASA/WMAP team wanted to write 'quantum fluctuations' rather than 'intricate, rich, precise order' on the label on their cosmology diagram was because they were taking the structure for granted and drawing attention to the small departures from smoothness, because these small departures were subsequently amplified by the action of gravity, so they can be thought of as seeds from which interesting things such as galaxies eventually would grow. So the departures from smoothness are significant, and when they are your main interest, you will want to draw attention to them. Please don't misunderstand me: I think the theoretical and experimental work in this area of science is impressive and rightly celebrated. But my purpose here is to try to 'see' the early universe in a full, balanced way, and to show it in full to others. The growth of galaxies, like the growth of hurricanes, is caused by much more than some initial small imbalance. It is caused by the exchange of energy and momentum amongst a complex collection of interacting quantum fields, whose billion-year evolution is described by physical laws of extreme precision and sensitivity. The combined effect of the balance of the initial state, and the regularity of the subsequent motion, is about as far from random as one can imagine, as I will now show.

Admittedly, when we discuss the earliest moments of the universe, we are entering unknown territory from a scientific as well as a philosophical point of view, as I underlined in Chapter 5. The physics of the very early universe is not known to us with any confidence, because conditions were almost certainly outside the regime in which our current knowledge of physics is applicable. Our attempts to understand this early era (the Planck era) are a work in progress; none are certain. But our knowledge of the universe which emerged from that period is more confident, and *everything we know about the universe which emerged from the Planck era asserts that it was very highly ordered.* After all, it has plenty of structure now and its entropy has only increased since early times. But a low entropy is, by definition, the absence of randomness and the presence of order.

Various authors have tried to make quantitative estimates of this. It is not a fully developed science, but a way of getting a better intuition about the early universe. Carroll and Tam, for example, present an analysis that can be used to estimate the degree to which the early universe was in a special, as opposed to an ordinary, state.[2] By observing the universe now, especially the

---

[2] Sean M. Carroll and Heywood Tam, 'Unitary Evolution and Cosmological Fine-Tuning', July 2010, 34 pp. CALT-68-2797; preprint arXiv:100.

faint background of microwaves and the distribution of galaxies, we can determine how smooth it was at the moment called 'recombination' when the background waves were emitted. This background (called *cosmic microwave background radiation* or CMB for short) is itself a fascinating piece of physics and a wonderful clue to the nature of the universe, because it is light that was emitted when hydrogen atoms formed out of protons and electrons about 14 billion years ago, so it tells us what the universe was like then. It turns out that the universe was very smooth. This is surprising because gravity tends to clump things together; the observed smoothness is a sign that the conditions were special. To find out how special, one can track the evolution back to earlier times using the mathematics provided by Einstein's general theory of relativity, and our knowledge of physics at high energies. We can thus extrapolate with reasonable confidence back as far as the moments when the temperature was at the 'merger of nuclear forces' boundary-post (10^{28} kelvin) described in Chapter 5. Then we can pose the question: of all possible configurations of the universe at this early merger period, what proportion of them were smooth enough to produce the universe as we observe it to be? Carrol and Tam calculate that a conservative estimate of this fraction is

$$\text{fraction(sufficiently smooth)} = 10^{-66\,000\,000}$$

That is, the decimal number 0.000...1 with sixty-six million zeros before the 1. It is hard to understand a number as small as that. It is mind-boggling. One struggles to find similes. But it makes it unquestionable that talk of the early universe as if it was largely random is misguided. The mental image of a mere sea of fluctuations is certainly misleading.

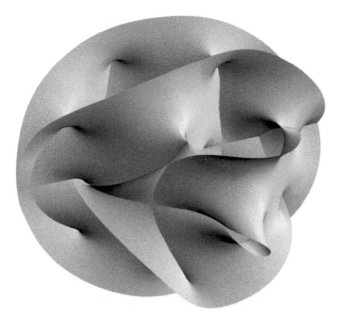

Figure 7.2. A projection into three-dimensional space of a Calabi–Yau manifold, an abstract mathematical object that appears in the mathematics of M-theory, which in turn is a theory of space, time and matter designed to model the quantum and gravitational effects observed in the universe.
Source: Wikimedia Commons.

In search of a better mental image, we do well to adopt methods to visualize the state of a set of quantum fields, including space–time itself. A rich set of mathematical ideas is involved, including, for example, a collection of concepts and methods at the meeting of physics and mathematics which has become known as M-theory. The study of M-theory invites one to contemplate multi-dimensional membranes with a rich set of possibilities for their curvature and topology. These are technical terms for how membranes can be shaped and can connect to themselves in multiple dimensions. These membranes represent space and time and quantum fields all together in one framework. We sometimes see beautiful pictures of three-dimensional projections of higher-dimensional structures called *Calabi–Yau manifolds* which arise in the study of M-theory. The pictures show bunches of petal-like folds curving back on one another and crossing over like high-dimensional origami figures with the creases all rounded into an elegant smoothness. Such pictures hint at the beauty of the structures embodied in the physical world from its earliest moments. Far from the seething mass of randomness which some physicists like to suggest, it was a multi-dimensional field in a beautiful, intricate and highly special configuration among all its possible configurations, and it evolved not randomly but in precise agreement with the complex set of equations that have come to be known as the Standard Model of particle physics. Rather than a swarm of random oscillations, it was and is a marvellously and intricately foliated multi-dimensional unfolding set of rose petals.

CHAPTER 8

WHAT SCIENCE CAN AND CANNOT DO

In the preceding chapters I have explored the beauties and marvels of the physics of the early universe, and made brief comments on the metaphysics. The combination of general relativity and quantum mechanics, called *quantum gravity*, is the fascinating edge of current physics, and it meshes in a truly insightful way with our observations of the large-scale nature of the universe as a whole—cosmology. It is this meeting which makes physicists want to wax lyrical about the inroads we have made to understanding the nature of the physical universe. But waxing lyrical must not be allowed to become waxing illogical. Physics cannot explain why there is anything physical, nor why physical stuff is deeply mathematical. No rational discourse can explain the very terms of discourse which it adopts and takes for granted.

By pointing this out, I am not intending to suggest that some other approach to metaphysics can be proved to be right. It is my present purpose only to say that quoting physical behaviours of physical things simply fails to gain any purchase whatsoever on certain basic metaphysical questions. I mean questions about how physical existence itself comes about or how anything invoked in physics is able to *be* as opposed to *not-be*. *Qua* physicists, the best we can do is to say, of the physical world and its existence, 'this is

Liberating Science. Andrew Steane, Oxford University Press. © Andrew Steane (2023).
DOI: 10.1093/oso/9780198878551.003.0008

what it is; this is its nature at the level of parts and their motions; this is what exists physically'.

This remains a beautiful and fascinating thing to say. It includes the chance to make statements about what the universe was like in the very early Planck era, for example, and to study the dynamics—dynamics which may well have involved a process like the one we considered in Chapter 5, involving a 'false vacuum' and consequently evolution like dew-drops condensing or a barrier opening. But when we do this we will be talking about the universe itself, in an early state: the state of the fields or the membranes or whatever they are. We will not be talking about something coming from nothing; we will be talking about something evolving from an earlier something, an earlier something already furnished with a mathematical description.

The very fact that it has a mathematical description implies that it is not nothing!

Stephen Hawking (1942–2018), the well-known mathematical physicist, was among those who, while rightly celebrating the progress and insights of science, went on to overstate what can be concluded. His *The Grand Design*, co-authored with Leonard Mlodinow (Bantam Books, 2010), sent perhaps a helpful prompt to philosophers to engage more fully with science, but also failed to achieve the level of philosophical correctness which is necessary for sound argument. He implied that physics can stand in for metaphysics. However, he also sometimes wrote more cogently about the situation, and one of his statements I very much like. It is a statement about the whole framework of fundamental physics which bears repeating:

> It is just a set of rules and equations. What is it that breathes fire into the equations and makes a universe for them to describe?

This is exactly right. When we make progress in theoretical physics we are making real progress, and coming to understand better the nature of the physical universe. But all our equations are just that: equations, mathematical statements. How can mathematics give birth to physics, to physical things? It cannot. It can only describe, and show links between one mathematical concept and another. That those same links should also be on show in the properties and behaviours of elements of the physical world remains a marvel that mathematics itself cannot either imply or account for. I can write equations till the cows come home, but the mere fact of having equations does not imply that there will be any cows, or any physical home for anything. A mathematical abstraction does not, in and of itself, imply anything at all about physical existence. The rules of chess do not themselves cause there to be chessboards and chesspieces.

I will now summarize the broader issue here; this will act as the conclusion of the argument of the book so far.

Physics is the name we give to the set of intellectual disciplines whereby we find out better and better descriptions of what is there in the physical world—of what has physical existence. None of those disciplines furnish tools to say how the physical world (space–time, quantum fields etc.) came to be there or comes to be there. As soon as we appeal to an idea based in general relativity or quantum field theory or some descendant of them such as string theory, we have *already* adopted a physical description. But a physical description is a description *of* something. Philosophers trained in metaphysics know this very well; it would be regarded as elementary. Some physicists seem to want to ignore it. The discussion of bubbles and virtual particles applied to space–time is a discussion of that which is already physical; that which already exists. Otherwise how can one assert and apply to it any mathematical equations at all?

One remains free to assert a belief that the physical universe has no prior or underlying cause and simply exists or came to exist for no particular reason, if that is what one thinks. But one may not properly call upon physics as a support for such an assertion.

From the perspective of physics, we can only say that the physical world exists and has the properties it does, including the properties on show in the Planck era (and before the Planck era, if there was a before). We might add: if there is a reason for this state of affairs then it is one that we are unable to find out by the method of studying general relativity or quantum field theory or any other physical theory, because physical theories are essentially descriptive.

Afterword

It is hard to come across as basically positive when your gardening efforts include a lot of pulling-up of weeds and cutting-down of brambles. For this reason I decided mostly to avoid a list of books or articles which might be interpreted as a list of culprits. But for illustration I will mention one exhibit, a book by expert physicist Lawrence Krauss with the title *A Universe from Nothing*, and, for the removal of all doubt, the subtitle *Why There Is Something Rather than Nothing*. The text of the book presents the kind of process which we have briefly surveyed in Chapter 5, but does so at greater length and with a good store of expertise and effort to help the general reader appreciate the physical and mathematical ideas. But the whole discussion is a prime example of using the word 'nothing' to refer to a specific and weighty *something* (the set of quantum fields in all their precision and glory, which amounts to the whole physical universe) and then claim afterwards that no misdirection was intended. What can one say to such sleight of hand? It is, I think, the product of a club, a group-think, rather than a lone individual. But the individual spokesperson carries sufficient of the responsibility that perhaps they can fairly be asked to help undo the damage.

I think the peer-reviewed scientific literature is mostly careful and proper in the details of what any given paper or calculation presents, but the headings and titles are often misleading, and invite a confusion between physics and metaphysics. This is even more true of off-the-cuff remarks which physicists sometimes include in lectures to a wider audience. Popular science writing in magazines, blogs and interviews has, I find, a low average standard in this area at the moment; hence the motivation for much of the first part of this book. It is important that the community of professional physicists does not behave in such a way that it becomes possible to accuse it of perpetrating a fraud on the general public.

SCIENCE, SCIENCE FICTION AND THE MULTIVERSE

In this and the next two chapters I will examine what kind of thinking is going on when we try to grapple with ultimate questions about the origins and nature of the physical world. In this chapter we consider the 'multiverse' idea.

What is the difference between science fact and intelligent science fiction? This question is more subtle than at first one might suppose, and I will not answer it fully. But an important element of the answer is that science, as opposed to a less restricted play of imagination, can shape claims that are, to use the technical word, *falsifiable*. This does not mean the claims will be proved to be false. It means that they venture to say something specific, something that can be tested by empirical investigation and might be found not to hold. So, for example, if someone makes the claim that certain electromagnetic phenomena can lead to a propulsion system with no need of fuel (such claims have been made), then it is a falsifiable claim because one can test it, either by checking the argument for correct use of established knowledge, or by building the proposed system and finding out if it works. On the other hand, if someone makes the claim that many significant choices made by men are the outcome of a subconscious desire to murder their father and marry their mother (another claim that has been made),

Liberating Science. Andrew Steane, Oxford University Press. © Andrew Steane (2023).
DOI: 10.1093/oso/9780198878551.003.0009

then unless a test can be proposed, it is not a falsifiable claim. If it is not a falsifiable claim, then whether or not it is true, it does not help us make progress, and we are probably well advised not to bother about it very much.

Modern physics has thrown up a bunch of claims which may or may not prove to be falsifiable. They include, for example, the claim that the universe we inhabit is just one among a vast array of further universes which are also physical, made of things like quantum fields, and which exist in other parts of a sort of huge hyper-space; such an array is called a 'multiverse'. This claim may be said to be falsifiable up to a point, in that if it is based on a theoretical framework such as a quantum theory of gravity, then one might argue that the tests which can be made on the theory are also tests of the multiverse idea. But someone might hold the multiverse idea even if the particular quantum gravity theory that suggested it proved to be wrong. So then, what is the test that would show the multiverse idea to be wrong if it is wrong? If there is not a test, then the idea is not falsifiable. That means the idea has little to offer. It can't be shown to be right, and nor can it be shown to be wrong if it is wrong. But that means it is not saying anything whose meaning can be pinned down and connected to other statements. Such an idea has little to contribute to further understanding. It amounts to an intelligent guess that seems to point to a fruitful line of scientific enquiry, but really just points to a combination of idle speculation and a never-ending sudoku puzzle.

Scientists will, and I think should, pursue this multiverse idea if they do so in hopes of finding empirical tests. For example, it may be that one part of the multiverse could impact on another in ways that would be observable to us. But there is also a judgement call that has to be made, to decide how far to push this—how much time and energy to spend on ideas so far removed from clear evidence or a clear sense that one knows what one is doing.

Multiverse ideas tend to be associated with a certain kind of situation in quantum field theory or string theory: the situation when the mathematical tools at work in the theoretical framework are powerful, in the sense that they can handle a wide variety of circumstances. The trouble is that when they become powerful in this sense, they thereby become power*less*: powerless to single out one circumstance from another and give reasons why one circumstance as opposed to the other actually happens. So, in a move which might fairly be called one of desperation, some physicists have declared that in such cases, all the circumstances happen. Never mind that we only detect one outcome, all the other outcomes have also happened in the past, or are also happening now, in some other region of the multiverse, it is claimed.

It may be so. Or maybe not. How would we tell?

One of the attractions of the multiverse idea is that it seems to offer a way of understanding how it came about that the universe we inhabit is so special. I mean, special in that our universe harbours life—conscious life, the life we experience. This is a special condition, arguably, because it requires very particular values for global physical properties such as the rate of expansion of the universe, the strength of gravity compared to electric forces, the entropy of the initial conditions, and other such things. This observation is called *fine tuning*. The way the multiverse idea seeks to address this is to say that if there are many universes, in which all possible laws of physics, or values of basic parameters, are given physical expression, then some will have the propensity to develop conscious life, and those are the ones in which conscious life will develop. On this view, it is suggested, no matter how special the circumstances are that required for life to arise, they will occur somewhere for no reason except that they are possible. The trouble with this is that it represents a refusal to think that there could be anything more to discover or any further understanding to gain. No matter how special the observed universe turns out to be, one just shrugs one's shoulders and declares 'well some universe or other had to come out like that'.

There is a kind of logical impasse here. In the past, if a proposed theory of physics made many central predictions which were found not to hold in laboratory observations, then we would conclude that the theory was simply wrong. Now we are contemplating a proposed physical theory—a multiverse hypothesis—which holds that physical stuff and the universe at large is, with extremely high probability, very different from what is observed, and with extremely low probability as it is observed to be. But we are not invited to conclude that the hypothesis is almost certainly mistaken; we are asked to trust that it is correct!

The trouble with this is that it is counter to the whole spirit and nature of scientific enquiry. The very nature of rational enquiry in the physical sciences involves effort to find ideas which correspond to the observed happenings, not some other hypothetical happenings which are not observed. Science also involves judgement about when it is appropriate to appeal to coincidence and announce, of some event, 'oh well, it just happened to come out like that'. Multiverse arguments urge us to make that kind of statement about basic properties of the observed universe. Proponents may judge that this kind of shrug of the shoulders is the best we can hope to do. But since this amounts to giving up on the attempt to understand further, I am reluctant to do it. Also, I consider that it is premature to do it now. We can probably gain further understanding. But the further understanding we need almost certainly will not come just from fundamental physics.

An interesting property of the multiverse concept is that it makes conscious life guaranteed to happen somewhere, once one takes the existence of the multiverse itself for granted. In the multiverse, life is not an accidental result; it is guaranteed. And there is no surprise that a universe with life in it has special properties—it is a self-fulfilling condition. But it is not clear whether this type of approach amounts to much more than guesswork. Also, the intellectual gain of this kind of 'solution' of the fine-tuning problem comes at a great cost: the cost of asserting that somewhere, in some other universe, there are the most terrible conditions of unrelenting injustice and pain. So we are invited into a thoroughly depressing consideration that the joys of some universes are only there at the expense of the horrors of others, unless there are further rules to prevent it. I call this a depressing consideration because I don't think the good places compensate for the bad ones. But a possibility that deserves our active interest is the possibility that there are profound constraints here, such that suffering is itself connected to, or only possible in the context of, the kind of transcendence which human suffering allows, as when a person disabled by injury goes on to become an athlete, or when a humble prisoner shares his bread with another, or as in the astounding example of Stephen Hawking's life with motor neurone disease, or when a person with a wasting disease manages to offer support in courage and comradeship to their loved ones, not letting the disease define them. Rather than making conjectures about multiple universes, treating them like so many numbers in a mathematical puzzle and adopting a mindset divorced from imagination or empathy, we can instead explore issues of meaning and pain and love which tell us what is involved in being a person in any kind of physical world.

Our own universe is quite painful enough. But the life we find among ourselves is of a very particular kind. It has, woven into it, all the subtleties of aesthetics and ethics, hope and the longing for justice, and consequently depth, value and meaning. It is a life worth living, I think. I also think we can make further inroads into getting a good all-round sense of this. But to speak of multiple universes purely on the basis of some mathematical equation is to embark on a kind of thinking which feels wrong to me. It feels too much like converting everything around us, including people, into an equation. I don't think the equations we currently explore in physics are anywhere near adequate to describe the universe as it actually is. I doubt that the very language of algorithms and equations of motion is adequate. But this is the only language that physics has learned to speak so far. I acknowledge that this doubt of mine is a subjective opinion, but I offer it here because it may help readers to know that one can hold such an opinion without thereby rejecting any of the wonderful discoveries of modern physics.

None of the above denies outright that there may be multiple universes. It denies merely that the apparatus of modern physics is able to give confident knowledge of such things.

In science we learn to pick the simplest line of explanation that fits the facts, but we admit that the word 'simple' is hard to define. One who does a lot of theoretical work may be drawn to simplicity or beauty in the equations, and not care how many universes are said to exist. One who spends time and effort wrestling with the concrete stuff of the world we inhabit, in an effort to get it to reveal its secrets, may be drawn to the simplicity of saying that there is just the one universe. After all, this statement is consistent with all the empirical evidence.

As I have said, different people will have different gut instincts about the multiverse line of argument. It is too speculative to be called science in the ordinary sense of the word, but it is not simply science fiction either. This is because it is an attempt to say what might be so, rather than an attempt to invent worlds in order to explore ideas about the nature of the human condition. My own gut instinct is that I am uneasy about 'conjuring into existence' (by a multiverse line of argument) myriads of universes where a vast quantity of suffering might be going on. I would rather hesitate, and ask what we can learn about the meaning of what happens in the universe we know.

The reader will notice that, when contemplating what we might come to think about the multiverse idea, my approach includes an awareness of pain and suffering, and a concern for meaning and value. When people have written about multiple universes in the name of science and shown no alertness to such issues, it illustrates the way science compartmentalizes thought. It also illustrates a modern tendency to emphasize types of analysis where everything is defined and captured. The multiverse concept might become more coherent and less arbitrary if it had properties which guaranteed that there was always available a response to suffering which allowed it to be transcended, to be wrestled into serving a good purpose. I don't think physics will ever furnish the tools to find out about that. But other areas of discourse might.

CHAPTER **10**

COULD IT SIMPLY BE?

When we contemplate the quantum fields and the early universe, we are working our way right up to the edge of physical existence itself. We peer over a metaphysical cliff and wonder whether physical stuff just *is*, and there is no accounting for why it is, or whether there is more to discover. Maybe there is a reason and a further cause, a non-physical one. The cause we are talking about here is not a cause that acts merely at an early time, but one that causes all of time. We are now using the word 'cause' in a metaphysical sense; a 'cause' that provides whatever is necessary to 'breathe fire into the equations', to use Hawking's metaphor, to sustain the bridge from mathematics to physics, from potential to actual.

The two possibilities are either that the universe just is—'it can simply be' as Hawking put it—or that the physical universe has a support which is not itself a further element of what is physical, but is an all-embracing *other*.

The first thing to say about these two possibilities is that we will not be able to use the tools of science and mathematics to weigh their relative plausibility or intellectual contribution. Science has the tools to discuss and describe that which has physical existence, but those tools cannot be brought to bear

Liberating Science. Andrew Steane, Oxford University Press. © Andrew Steane (2023).
DOI: 10.1093/oso/9780198878551.003.0010

on the question of physical existence itself (as I have already argued in earlier chapters).

In order to make a little progress with this issue, one can ask, of any given item, whether it is possible to conceive that that item might not have come to have existence. For example, one can easily imagine a world which has no dodos in it, because we live in such a world. The dodo is a species of flightless bird which once inhabited Mauritius, but is now extinct. Clearly, then, it is possible to conceive of a world with dodos, and it is possible to conceive of a world without dodos. We can say, therefore, that dodos are *contingent*—they don't have to exist; their existence depends on, is contingent on, other things, such as natural history and the behaviour of predators such as humans.

When we examine larger things such as planet Earth, or the solar system, or the Milky Way galaxy, they too appear to be contingent. One can easily imagine a physical universe which did not have them. Pushing this line of thinking further still, it appears that the whole physical cosmos is itself contingent. That is not something we can claim to know for certain—we are here pushing up against the boundaries of human knowledge—but it seems reasonable to say that the physical universe does not exist by logical necessity. One can conceive of a case where the universe we inhabit never came about. A case where its quantum fields or strings or twistors or whatever simply are not and never were. In that case we would not be here to wonder about it, of course. But there does not appear to be any sequence in logic whereby one can arrive at physical existence from pure reasoning or from pure mathematics. Therefore physical existence is contingent: it is dependent on some other consideration which we hardly know how to even express in words—some sort of support or quality to make it possible that physical existence can happen rather than not happen. Equally, it is logically possible that some other type of universe would exist with inhabitants able to ask such questions, and those inhabitants might find that their universe also was contingent.

So then we are driven to ask the question: can there be something whose existence is not contingent? Something which cannot fail to be? Again, this is not an area of certain knowledge, but there is a broad consensus that the nature of a non-contingent item would be different from the sort of nature that physical items have. It would have an existence not like physical existence, but more like the existence which the truths of mathematics have, and which the truths of interpersonal relations have. I mean for example the truth that $2 + 2$ is equal to $1 + 1 + 1 + 1$, and the truth that kindness is better (more constructive, creative, imaginative) than bitterness, and that education is better than propaganda, forgiveness than revenge, and so on. Such truths are non-contingent because one cannot conceive of matters being otherwise. It does not make any sense to suppose that $2 + 2$ is not equal to $1 + 1 + 1 + 1$ in some other world,[1] and it does not make any sense to suppose that revenge, propaganda and bitterness are best in some other world. So the non-contingent includes a collection of truths such as these.

That is, approximately, how far one can get with pure reasoning. We still have not settled whether or not the physical universe could simply be, with no further (non-physical) cause or support, but we have at least glimpsed the idea that there does not need to be an infinite regress here. We can understand that questions

[1] I am taking it for granted that the mathematical symbols adopted here have their standard meanings.

of existence can come to a finish at some sort of non-contingent base. We can also, by pure reasoning, make plausible conjectures concerning some properties of such a base. But our powers are limited. Just as we need empirical methods to understand the natural world, we will need a more empirical, interactive approach to learn about the non-contingent.

But this empirical approach challenges us deeply. Here we encounter a challenge that goes to the heart of who we are. Questions of this kind are not going to be solved for us purely through a process of deduction by reasoned argument. It is almost certainly true that there is no cautious or self-evident premise from which you can start out, and then, purely by a process of reasoning, conclude that the physical universe can simply be, nor that it cannot simply be. The question is not going to be solved for any of us by some argument like that. The question can only be solved, for each of us, by bringing in further types of consideration. To engage a question like this one we have to try to plumb the depths of our existence using all the apparatus at our disposal, and some of that apparatus is deeply personal.

It is a very widespread feature of human life that people, upon making some effort to plumb the depths of their experience of the present moment, find themselves drawn to responses such as gratitude and moral objection—gratitude for beautiful, fulfilling things; moral objection to ugly, unjust things. They find themselves 'swimming', as it were, not just in a bunch of physical realities but also in a sea of aesthetic and moral principles.

If we want to say, of the physical world, 'it can simply be', then we may also be inclined to say 'it can simply be' when it comes to a moral principle such as the principle that nurture is better than torture. We might want to say that friendship is better than feuding, and that violence should be a last not a first resort, and so on. We thus find quite a large collection of things we may care about, and it is interesting to note that they don't present themselves as many disparate things, but as different aspects of a single whole. So the 'it can simply be' which one was inclined to say of the physical universe turns out to be a much richer statement than one realized—the *being* of the universe includes that it enables all these moral principles to be enacted and lived by. The world's physical nature is not described by just any old bunch of mathematical equations; it is a very particular bunch that allows this richness to be bodied forth. But how can mathematical equations know about justice?

With such a question I can set off any amount of philosophical reaction and reflection, but it is not my purpose here to devote space to that. What I want to do is communicate a piece of information. It is information for anyone who has been inclined to say 'it can simply be' of the various values they may have. When we add up sufficiently many of these 'it can simply be's we begin to see, dimly and tentatively, the notion of that which is *non-contingent and supremely valuable*, a beautiful unity and fullness which draws our commitment and is not centred in ourself. This is a hint of what can be meant by the word 'God' when the word is used carefully. But be careful. That same word is also now widely used for another kind of thing, a sort of super-powerful being, a bit like the mythological beings invoked in ancient Greek or Norse or Hindu religious mythology and called 'gods'.

Those stories explore aspects of human experience and psychology in an interesting way, and they touch indirectly on deep aspects of what it is to be human, to have hopes and pains, tri-umphs, betrayals, bewilderment. But such stories are also 'safe', because we can distance ourselves from them. To the modern mind they are clearly fictional and they leave us in charge of ourselves, free to come to our own conclusions about what is real and what is not. But the task of human life involves going further than this. To be alive is to be in life—to be part of a network of relationships—to venture our very selves in a relation which we do not define but which defines us. When approaching the non-contingent—those aspects of reality which cannot fail to be—we are facing a challenge, but a very creative one. Instead of assessing stories or exchanging ideas, we are opening ourselves to the very reality which has given rise to us. We find ourselves in territory where our attitudes and habits are tested, and they may not measure up.

When approaching the most supremely valuable and beautiful, all we can do is admit our need to learn. Our task is not an intel-lectual one of deciding whether or not something exists (a task which somehow manages to be petty and pompous at the same time). Our task is to pay attention to what in reality defines who we are and who we should aspire to become. That is what the word 'God' has often meant, in the English language. If that is not what it means to you then perhaps another word may help, or a collec-tion of words such as love, and justice, and beauty. But beware of making your collection too tame. If it leaves you in charge then you have not broken free.

CHAPTER 11

RELIGIOUS IMAGERY

Imagine a child asking their parents the age-old question 'where do I come from?' Suppose a parent takes a moment, smiles a warm smile, and says, 'well, what can I say? You come from a great mystery. You are a child of hope, and a child of willingness, and also, sad to say, a bit held back by our faults and frustrations, but deepest down, I want you to know, you are a child of love.' Now suppose that same youngster goes to school and there they ask the same question, and the teacher replies, 'well, your mother's body created a tiny living cell, and your father's body provided an even tinier swimming cell, and these two came together, and inside your mother's body this special cell split into a pair, and these grew and split in their turn, and more and more cells were thus produced and your body grew.'

So far so good. Both these replies have told their truth and I value both. But now suppose a man came along and announced that the biological reply *replaces* and logically *refutes* the spiritual reply? What can one say to such a man? Where do you start? He is just an ignoramus, a fellow of the society of royal idiots.

But I have read many statements displaying this kind of ignorance. It has become commonplace to say that in its work on cosmology science is now tackling questions which were previously the purview of religion, and making a better job of

Liberating Science. Andrew Steane, Oxford University Press. © Andrew Steane (2023).
DOI: 10.1093/oso/9780198878551.003.0011

answering them. This kind of claim commits a category error. That is a polite way of saying that the claim is nonsense.

Any ordinary non-religious person who owns a spade and a screwdriver, and who has a couple of household jobs to do, will use the spade to dig the garden and the screwdriver to tighten a screw. What about the religious person? Will they get in a muddle and use the screwdriver to dig the garden and the spade to tighten a screw? Of course not. They will use the tool appropriate to the job in hand, like anyone else who is not a complete idiot. So if they wish to find out what the universe was like in its earliest moments, the person with religious commitments will use the techniques of physics and astronomy. And if they wish to find out how to become a better person, they will use the practices that have proved their merit in helping with that.

One reason why it is nonsensical to contrast a well-researched scientific observation about origins with a high-quality religious statement about origins is because science is itself the tool which good religion adopts for purposes of study of the nature of the physical world. High-quality religious response does not consist in abandoning the tool or ignoring what it shows, but in joining a communal effort to interpret what it shows.

So my first point here is one about science and the information it gives. That information should not be interpreted as a sort of alternative religion, nor an alternative to religion. It is simply part of what we can all learn. We can all learn about the history and properties of the natural world. If we learn, then we will learn that its properties long ago were the ones that can be deduced from the physical evidence—the properties that are discoverable by astrophysics and particle physics working together. Science belongs to everyone.

The part of our human nature that is called religious, or that religious practices make contact with, is concerned with synthesizing or 'reading' the information that science provides, along with the information to be obtained from ethics, art, everyday life, human encounters, history and silent contemplation. It is not something that science replaces, but something that science informs.

I mentioned above the notion of 'a high-quality religious statement about origins'. An example of a high-quality statement is the notion of *creation as free gift*. The notion that there is, underpinning the very fact of physical existence, a quality of *gift*, of that which is freely given, so that it is appropriate to see our own role as neither fatalistic submission nor self-centred choice but as an act of reciprocal generosity.

In response to religious impulses people have found it appropriate to create figurative stories and works of art, among other things. These works can offer a 'way in' to showing what it is that religious practices offer. But I have noticed in the modern scientific community, or at least the part of it that I have encountered, a strange sort of refusal to allow this. It can take the form, for example, of refusing to treat artworks on anything but a surface or literalistic level. It is the rough equivalent of dismissing a painting such as Picasso's *Guernica* on the argument that it does not show the anatomy of a horse correctly. Now typically people do not treat that particular artwork so ignorantly, because they are willing to believe that the artist knows what he is doing and they understand that art does not always adopt, or function best, in a literal, realist style. But faced with religious artworks, it has become common for people to refuse to adopt anything but a literal or realist interpretation, and then dismiss the message they imagine the artwork conveys.

In order to illustrate this, I shall use as an example the painting by Michelangelo on the roof of the Sistine Chapel, the one commonly known as 'The Creation of Adam'. It has become one of the most widely recognized paintings in the world. I have picked Michelangelo's painting for discussion here because it was used in a public science lecture at Oxford University in which the lecturer claimed that modern physics refuted it.

The question I want to ask is, what did the artist, Michelangelo, and the patron, Pope Julius II, intend to communicate by this work? Or, perhaps a better way of putting it, what is the message conveyed in the work itself, as the artwork that it is? What is it communicating to us, the viewers? What does it invite us to perceive, to realize, to think?

The image concerns humanity and God. And there we have the most disputed word in the English language, and in any language. To read an image or a story concerned with God, one has to begin by adopting a certain discipline, which is the discipline not to assume that you already know what the word means. One must shelve one's preconceptions, and not allow them to occupy the foreground. One must admit that one is entering an area of mystery, in which one person communicates to another something of what they perceive about our situation in its foundations and in its hope and possibilities.

The image painted by Michelangelo shows the man, Adam, from the story of the Garden of Eden in the early part of the Bible, and a representation of God in the form of a man, reaching to Adam, their fingers close but not quite touching. The figure representing God is accompanied by twelve other figures, including a striking feminine figure in the centre of the group, leaning in a way which provides a balancing counterpoise to the reaching-out motion, and the whole image is very dynamic.

The artistic power of this painting is unquestioned; its brilliance of composition and its accomplished portrayal of muscular human anatomy is easy to appreciate. But it is plainly a painting with a religious message; what message might that be? Is it Michelangelo's desire, for example, to suggest to us that God is like a big man in the sky, a bearded white male one to be specific? Or is it his wish that we should think that the first human being was created abruptly, in a sudden miraculous event? I would like to suggest that neither of those ideas are part of Michelangelo's main purpose. Or, to be more precise, neither of those ideas are central to what the work stands for in and of itself. They are irrelevant distractions. The big message of the painting is in its overall composition, in its sense of movement and vitality, in the passive pose ascribed to Adam and the active one ascribed to God, in the dignity of both forms, and the work celebrates them, along with the others around God, especially the woman, who has a strong presence in the image. The message is in the sense of friendship which pervades all the relationships on show. When the artist used a human form for God, and a white male one, he is merely using a convenient shorthand, a kind of musical notation; he assumes we are all grown-up enough to know that this element of the painting is figurative and does not contain the message.[1] The intention is to show us something about the nature of God not in what God 'looks like' but through what

---

[1] Michelangelo was an intelligent man, and deeply thoughtful about his art. I don't know exactly what his religious commitments were, but for sure he knew that religious matters are subtle and religious symbolism is just that—symbolic, indirect, and capable of being misread.

God is *doing*. Michelangelo wanted to paint *uncontainable vitality*; *passionate creativity*; *fellow-feeling*; *active reaching out*. But how do you paint uncontainable vitality; passionate creativity; fellow-feeling; and active reaching out? Well, one way is to use human figures in poses and gestures that convey those things.

Now I would like to acknowledge that in the context of particular kinds of injustice, such as the injustice that has been done to dark-skinned people by light-skinned people, one would paint a different artwork, one designed to approach such a context thoughtfully. Also, this artwork, like any good thing, could be used for bad purposes which were never the artist's intention. But when we are being fair to the artwork we will ask ourselves what it is doing with the issues to which it is in fact addressed. And when we do that then issues of ethnicity and gender are irrelevant. To be clear, I am not here denying that those issues have shaped our history, from which we need to learn and, to some extent, escape. What I am doing rather is saying that once we have got past that history, and learned from it, then we gain the freedom to look back and receive from any work of art what it can say when liberated from whatever untruth got mixed up in it. When you are looking for what truth is on show in an artwork then the attention has to be of a discerning kind; one which quickly moves on past possible untruths that the work may suggest, not distracted by them, but eager for what was really driving the imagination of the artist, the message which resonates in the heart as deep calls to deep.

It may be that some readers find it hard to think that the message could be anything other than blatant patriarchalism, or an outdated attempt at science, or ignorant guesswork in the absence of science. If you are such a reader then I would like to encourage you to step back a little and not leap to conclusions. Generations of viewers have quickly appreciated that what Michelangelo shows us here is *an act of generosity*. If you are seeing something else, then give some thought to why that might be.

On this danger of leaping to conclusions, a very helpful contribution is the work of the psychiatrist, neuroscientist and philosophical thinker Iain McGilchrist. In his books *The Master and his Emissary* and *The Matter With Things*, McGilchrist presents a large body of work in psychiatry, neuroscience, science more generally, philosophy and history. He is interested to uncover and weigh the different modes of perception that humans employ as they encounter the world. His work draws on a large number of studies in psychiatry and neuroscience, along with other areas, and it is focused on discovering what is the most insightful and least misleading way in which we are able to perceive the world and our place in it. This study is helpful because it provides a large amount of evidence that we have two types of apparatus in our brains for thinking, each capable of framing conclusions, and one of them is extremely poor at handling metaphor. When faced with symbol and metaphor, the type of thought process which we use in analysis and dissection has a tendency to latch on to a literal or familiar interpretation and then come up with a story to rationalize that interpretation. And furthermore, this part of our brain apparatus (a mainly left-hemisphere part) can remain confident of its conclusions even when they are plainly wrong. McGilchrist offers many striking examples. The plainly wrong conclusion in the present example is that Michelangelo was engaged in some kind of literal depiction of an ancient event. In order to know that such a conclusion is plainly wrong we do not need any hindsight or any appeal to modern science; we discern it because Michelangelo, as one artist to another, could immediately appreciate that the story of the Garden of Eden, with its snake and its trees, is itself a symbolic story. In order to interpret both this story and Michelangelo's art we need the other half of our brains in order to avoid deluding ourselves, the half which is good at understanding stories in their full context: the context of human life in all its richness.

In the list which I proposed above, of items close to Michelangelo's purpose, or of moods which the work itself conveys, I mentioned *fellow-feeling*. One of the noteworthy features of *The Creation of Adam* is that God is not shown in an imperial pose in formal imperial dress, which was a common practice previously in officially sponsored western Christian art. God is shown informally, loosely clothed, and consequently a lot closer in spirit to the everyday person. Also, as the reclining figure of Adam and the sweeping figure of God look towards each other the one's face suggests a gentle, open love, ready to learn; the other's suggests a focused, determined love, ready to teach. The genius of this artwork is in the way it conveys what moments of inspiration and comradeship are like. Although we work hard to learn, we also have moments when something dawns upon us, or opens up before us, a moment when we are passive receivers of the creative inspiration that enables us to be and to grow. Our growth in character and virtue is especially like this. Michelangelo presents that inspiring touch not as an overwhelming, invasive force, but as a spark that respects our independence, that seeks to create not a doll or a puppet or an obedient robot but a reciprocal movement of love.

The discoveries of modern and future physics and biology will not replace this, nor could they. What they are discovering is the physical process whereby the truth on show in the religious artwork came to be enacted in the world. It came to be enacted through a long sequence of events; and it has not finished being enacted. It is ongoing now.

Spark Gap
What do you think you are painting there,
Michelangelo, with your brush?
Is it a god? I hope not.
And I think not, too.
Sweeping towards the dozy Adam
in urgent, wakeful artistry,
yet holding back, the spark
about to quicken, not invade,
I see
not a bearded white male,
ungenerous cliché of a willed ignorance,
but an urgent, wakeful artistry,
sweeping towards the dozy Adam,
yet holding back,
the spark about to quicken,
not invade.

CHAPTER 12

SINKING THE SELFISH GENE

We now turn to evolutionary biology. I begin with some critique of Richard Dawkins's book *The Selfish Gene*. Then in
the next few chapters I discuss the public presentation of this area
of science, and I remark on the nature of the story of life on Earth.

I pick *The Selfish Gene* for critical comment because it has been
widely celebrated, so I am critiquing a widely diffused set of contemporary assumptions, not just one book or one author. But it
helps to name a book in order to be specific. When I first read
The Selfish Gene, many years ago, I was interested in the examples
drawn upon, and in the logic about the way the gene pool operates. But I also felt that there was something wrong with the overall
message. I felt that the wool was being pulled over my eyes.

The issue here is not whether the neo-Darwinian process really
happened; the evidence is that it did happen (though not quite as
simply as proposed in models developed in the first half of the
twentieth century). But this was discovered by other people, it is
not the distinctive contribution promoted by Richard Dawkins in
The Selfish Gene. The distinctive contribution is the emphasis on the
individual gene and the line Dawkins takes about the meaning of
it all. He proposes that the overall message is that living things are
gene-survival machines: that is their purpose and role, and that
is how they are best seen or most insightfully seen. This tells us

Liberating Science. Andrew Steane, Oxford University Press. © Andrew Steane (2023).
DOI: 10.1093/oso/9780198878551.003.0012

what we are, the argument goes. We don't need to be in any doubt or confusion about that any more, because in 1859 it was at last laid bare. Dawkins then gives a brief comment to modify this conclusion in the case of human beings, by appealing to the role of memory and cultural history.

After finishing the book I was still inclined to think that the truth not just of human beings, but also of other living things, is not chiefly that they are gene-propagators. I was inclined to think that the role and purpose of living things is much more insightfully seen as a combination of many features, functioning at many levels, so that when trees in forests grow tall, for example, it is not a mere waste of effort. So if I was right then there must be something wrong with Dawkins's approach.

It took me a long time to figure out what the problem was. It required one to think hard about how scientific explanation works and what does and does not follow from the physical mechanisms on show in the natural world. The fact that I could only do this slowly suggests either that I am slow or that the problem is difficult. I think I am in some respects slow, but this is not the whole situation. Perhaps unravelling this intellectual problem was difficult, by some objective standard of difficulty. But the fact that there was a problem, a failure of logic, was evident from the start. The logic that fails here is the idea that one can deduce the purpose of living things (if they have one) from the mechanism of the Darwinian process.[1] One cannot do that; it is illogical. The purpose of living things (if they have one) is found out by other means, and it almost certainly has very little to do with their propensity to pass on genes to their offspring.

---

[1] I use 'Darwinian' as a shorthand; it can be taken to mean 'neo-Darwinian' for the purposes of this book.

But why is this still a live issue? It remains contemporary because I was disheartened to learn, a few years ago, that *The Selfish Gene* is regarded by many scientists as the best piece of science writing or popular presentation of science that they have read. Why would I be raining on this parade? Why disheartened? I was disheartened because this suggests that there is an ongoing problem in the scientific community. Many scientists have simply been carried along by Dawkins's lively and pithy writing without noticing that it is correct in some respects and illogical in others. This suggests a readiness to be duped by eloquence; an unwillingness to adopt the discipline of sceptical enquiry. A second reason why this is contemporary is that it is an example of a wider phenomenon, a form of writing in the popularization of science which repeatedly adopts a specific *non-sequitur* (i.e. a bad argument; a conclusion which does not follow): the non-sequitur that since Darwinian evolution happened, 'therefore' there is no love or creative long-term reason for hope at the heart of reality or of what makes us who we are. There is no valid 'therefore' in such statements. The rest of this book will unpack this.

What one can be rightly sceptical of, regarding the argument of *The Selfish Gene*, is the philosophical claims being carried along for a ride beside the scientific ones, without being supported by them. For example, it is repeatedly implied or assumed that one can determine the purpose of an element of the physical world by finding out how that element came to be there. So, for example, if white-looking fur on a polar bear came to be there because previous bears with genes for black fur did not prosper and reproduce, then in *The Selfish Gene* we are invited to the conclusion that the polar bear's purpose and role is to be a multiplier and propagator of white-fur genes and other such polar bear genes. Put bluntly like that, this seems a bit of a leap, because it introduces teleology—the language of purpose—into, or tacked onto the end of, an argument that was about physical causes. The 'argument' being furnished to encourage us to see polar bears this way is not a logical argument. It is a piece of passion, an *O say, can you see, by the dawn's early light*, as we are invited to Dawkins's *land of the free and home of the brave*. Well I took a walk in that land of lumbering robots and found I did not want to live there, and since reason did not require me to live there, I decided not to.

Eventually I wrote a book which, while also discussing other things, refutes a major aspect of the argument put forward in *The Selfish Gene* and another popular work by the same author, *The Blind Watchmaker*. This was not to refute the standard evolutionary biology which indeed Dawkins presents eloquently, but to refute the claim that evolution is largely directionless. That claim is not scientifically established, and indeed such a claim is sufficiently far from the truth as to be essentially misleading. But, as I already said, one should not blame Dawkins alone; it is a wider cultural atmosphere that is to blame here. This is an example of the readiness of groups of people, at any given time and place, to resist some ways of thinking and welcome others. When something sits comfortably with cultural fashion, people are more ready to buy it without stopping to question with persistence and care. And when there is a background of cultural clashes such as the one which has played out in public education in America, then clarity may be even harder to achieve.

Previously some philosophers have detected a too-loose use of language in Dawkins's work, and critiqued it on those grounds, but this reaction did not get to grips with the details of the argument about genes and replication, and it has not reached a wide audience. The idea that we are puppets of our genes, and that our moral and aesthetic judgements are merely the lingering effect of mutations which enhanced reproduction long ago, is now widely thought to be a 'scientific' idea, or 'what science says'. But (as I will argue in this and following chapters) this is neither scientific nor what science says. So we do need fresh efforts to help the wider audience appreciate this.

In my own previous contribution I aimed to tease apart the mechanism of evolution, presented largely correctly in Dawkins's work, from the claims about the meaning or purpose of living things which he also makes but which are not scientifically supported. That doesn't mean one may not make them; it is just that one should allow the reader to see where the objective description finishes and the subjective interpretation begins. In the present book I will summarize this, and also bring out more fully the role of constraints on evolution. The situation might be compared to studying the flow of a river. A description of Darwinian evolution focused on genes is like a description of a river focused on water molecules. Each molecule just shoves its neighbours, so such a description does not get much of a sense of the overall direction of the large-scale flow. It seems as if it could be going anywhere, and since the whole river is molecules, it seems as if there is nothing further to add: the whole truth has been told. But this conclusion fails to see the effect of the river banks: the flow is in fact highly directional. In a similar way, the jostling of genes in the gene pool can appear directionless if one never steps back to take into account larger-scale principles which everywhere shape the natural world.

My claim to 'sink' or riposte the argument of *The Selfish Gene* is not, let's be clear, an attempt to overturn well-established facts about the evolutionary process. I share the general sense that Dawkins's writing gets to grips with that process insightfully and intelligently in many respects. Perhaps this is what others are affirming when they celebrate the book. But I think it is unquestionable that the book attempts to persuade us that gene replication and selection is the main story of life on Earth, and gene propagation amounts to the chief function or chief role or primary purpose of organisms in the natural world (their 'sole reason for living'). It is also implied that the interplay between negotiation, competition and cooperation which goes on in almost all of life is most insightfully seen and understood as entirely competition in varying degrees of directness. It is further claimed that organisms are 'slaves' of the controls placed on them by their genes. It is these claims that I refute.

For clarity in the following, let's label these areas in which claims are being made:

(A) purpose; chief function or role
(B) competition, cooperation and negotiation
(C) control.

First let us deal with item A: the notion of *purpose*. The mechanism whereby life has developed in all its varieties is the one outlined in standard evolutionary biology, which describes a sequence of cause and effect. It is not logical to infer from this that the purpose of a polar bear is to propagate polar bear genes, because purpose or teleology does not follow from causation. Nor can one justify a claim that reproduction is the most important thing a polar bear does. If one is telling a story in which genes are the protagonists, then reproduction suddenly looms large in significance—it becomes one of the most crucial elements of the plot—but this is just a fable, a dramatic device to engage our emotions. When you are talking about replicators, you will focus on the fascinating way in which the Darwinian process allows replicators to multiply; that will be the story you tell. It does not follow that that is the main story or the primary significance of what has happened, any more than the generation of heat is the main outcome of chemistry in the human brain. And it does not follow that the purpose of the bear is to be a gene-replicator. In fact, no single area of science can reveal to us whether or not a polar bear has a purpose in the grand scheme of things, nor what that purpose is if it does.

This logical observation does not on its own get to grips with the emotional power of Dawkins's approach. One might feel that Dawkins's extended metaphor, and the fable that accompanies it, are showing us the main theme 'really'. One might judge that the genes and the way they influence their own probability of replication are the most significant item, the central element in the explanation for life on Earth. But such a judgement can be refuted. To understand what is wrong with it one must take time to consider the whole nature of scientific explanation. The structure of scientific explanation is a much richer structure than Dawkins realizes. It involves multiple levels and multiple languages, and these languages or levels do not replace one another but inform one another. No one level gets the dubious accolade of 'central' or 'most important'.

Here is an example from physics. In order to understand the outcome of a collision of two particles, one can either study the forces which the particles exert on one another and the accelerations that result, or, instead, one can assert the conservation of momentum and the conservation of energy. The reader needn't know exactly how this second method works; the point is merely that it is an equally valid method and it does not need to track the details of what happened during the collision, because it appeals to general properties of the forces which are known to hold at all times. In biology, one can make similar statements about the structure of plants and animals—there are many features for which one does not need to track the details of what happened during cycles of reproduction, because the structures and behaviours that result have to conform to global principles such as conservation laws, thermodynamics, engineering principles, reaction rates and the patterns of social dynamics. In short, a vast amount of what is going on in biology has nothing to do with the genes as such.

Let me repeat that. *A vast amount of what goes on in biology has nothing to do with genes.*

No amount of genetic engineering will produce an organism which does not conserve momentum, and respect thermodynamics, and survive without adopting survival strategies that work. It is only after one has taken the rest of science for granted that one can announce that the replication of genes is what 'Really Matters' in biological evolution.

Let's compare the situation with claims that one might make in the name of chemistry. Someone might say that chemical reactions and the increase of entropy are the chief element in the scientific explanation of life on Earth. But to say that a mouse is a chemical factory, and so is a cactus plant, is to make one point while missing another. It is a true statement, but not one which helps us to understand what it is that distinguishes a mouse from a cactus. Similarly, the assertion that both a mouse and a cactus have a Darwinian history behind them is a truism, but not one which helps us to understand what distinguishes them from one another. To understand what is going on in the differing form and behaviour of mouse and cactus one has to look into many interesting physical, chemical, engineering, biological and social patterns and structures which have nothing to do with genes and natural selection *per se*. I really mean *nothing*. The conservation of energy is no more a result of natural selection than is Pythagoras' theorem. The same can be said for more subtle things such as the value of trust and mutual recognition in a social species with enough cognitive faculty to make such things possible. Please don't misunderstand me: trust has 'fitness' or survival value, just as other capacities such as efficient lungs and the ability to count do, and it will be amplified in populations which discover it. My point is that the *nature* of trust, what trust *is*, is not about genes and selection. The evolutionary process could not change what it is by one iota. Rather, the nature of trust, along with all the other principles (from physics, maths, chemistry, social dynamics etc.), are constraints within which natural selection operates. The ultimate rationale for the existence of all the various living forms is found partly in gene replication, but more largely in all the principles which are at work in each of these areas of science and of the natural world. Those principles shape the space that the mechanism of evolution flowed into. The narrative of *The Selfish Gene* has not noticed this and consequently is deeply flawed.

That concludes item A in my list. It is the main item and its refutation is sufficient for my purpose in this chapter. But let's look at items B and C as well. Item B is the proposition that in order to get a good insight into zoology and evolutionary history, the metaphor of competition is not just useful and central but overriding. By 'overriding' I mean the proposition that whenever we see negotiation or cooperation we should immediately look for signs of competition which has led to it, and then we shall have arrived at 'the truth'. For example, the mathematics of replicators such as genes (accepting, for the sake of argument, a simple picture of genes) is such that it can furnish a mechanism whereby cooperative behaviours of organisms can come about, because such behaviours enhance evolutionary fitness. Now I agree with Dawkins, and others before him, that tracing the causal process from 'selfish' replicating molecules to cooperative organisms provides a good and helpful insight into the natural world. But it does not follow that the cooperation is secondary or illusory and evolution is 'really' all about struggle and competition. Evolution is what it is, and it is largely about negotiation. That is, there is a deeply built-in tendency towards negotiation in the world of living things which change and inherit broadly as Darwin perceived. It is built in not by any mysterious life force or anything like that. It is built in by ontology: by the very way things are or can be. This is because any collection of parts which negotiate with their environment, and with one another, and thus arrive at complementary roles, will fare better than a collection of parts where each only seeks its own advantage in competition with the others. This fact is already descriptive of the processes and structures inside any single cell, and it also led to multi-celled organisms, and it is at work in relations between organisms. It is a principle which is as deeply woven into evolution as any principle of competition for resources and elimination of the weak.

The Selfish Gene might be compared to a book about integers which only ever mentions odd numbers. There would be plenty of interesting things to say in such a book. One could mention that any integer power of an odd number is odd. One could mention that products of odd numbers are odd, and their prime factors are odd. The number of Platonic solids is odd. And so on. Now suppose that such a book claimed to be teaching us about integers in general. Would such a claim be a half-truth (since half the integers are even)?[2] No: it would be simply untrue. It is untrue that all integers are odd. And it is untrue that all biology is competition.

The further claim advocated in the *The Selfish Gene*, item C in my list, is made through repeated appeal to concepts of slavery and tyranny. This claim is overstated. Indeed it is massively overstated. One might read this as an innocent example of the use of hyperbole: a deliberate exaggeration so as to press home a point, in which the reader is expected to understand that this is what is going on. But the whole basis of *The Selfish Gene* assumes that the relationship between genome and cell is one of controller to controlled. The model of cell management is assumed to be roughly comparable to the kind of centralized system of management which has been attempted in communist countries. But this is incorrect. What actually happens in living cells, and in whole organisms, is far from this. It is sufficiently far that the model assumed in *The Selfish Gene* is at best half true, or gives approximately half the picture. It follows that it is very misleading to present it as if it were the whole picture, and indeed much of modern discourse has been duly misled.[3]

---

[2] To be precise, the proportion of integers between 0 and N which are even tends to one half in the limit $N \to \infty$.

[3] Here is an example. A few years ago in a television documentary, the widely respected broadcaster and natural historian Sir David Attenborough presented a survey of his life's work with the BBC. In a clip from an early example, we hear him as a younger man marvelling on the touching attentiveness shown by a pair of monkeys to one another as they engaged in a lengthy and peaceful grooming session, sitting on a branch of a tree. The older Attenborough then added a disparaging remark on this commentary, saying that it had been 'hopelessly anthropomorphic'. The same broadcast alluded directly to Dawkins's

Cells are controlled up to a point by their genes, but there is no one-to-one relationship between genome and cell behaviour. The same gene sequences can code for different proteins depending on context; the same genome leads to cells of all different types in a given organism, depending on the cell environment. And as far as we can tell, it is not true to say that the DNA on its own accounts for this observation, at one step removed. The process is not quite like that. Rather, there is a two-way relationship of great subtlety between the DNA and the whole rich process of life and biosphere over the generations. Some of the cellular memory is not in the genome at all. And as has been widely remarked, DNA itself is chemically inert. Far from being the controller, it can equally well be seen as a resource which the cell uses. Similar statements apply to whole organisms. They are controlled and constrained up to a point by their own internal structures, but as one considers more and more complex forms of organism, one encounters more and more complex forms of social interaction in which the order of the day is *negotiation*. The chemistry and biology, including the genetic component, confers an *ability to negotiate* the environment, including the social environment, and this is not about control, tyranny and slavery. It is about freedom (within boundaries).

So much for *The Selfish Gene*. Now for the other book.

I claimed, earlier in this chapter, to have refuted a major argument in both *The Selfish Gene* and another of Dawkins's contributions: *The Blind Watchmaker*. Since this is a strong claim, I have been specific, above, concerning the first of these, and now I need to be specific about the second.

The Selfish Gene. What I find interesting here is that the older Sir David felt the need to distance himself from his earlier observation, whereas in fact that earlier observation had been entirely warranted, balanced and correct.

A major argument of *The Blind Watchmaker*, perhaps the central argument, is that biological evolution is unguided. One can admit the validity of such a claim up to a point; I don't want to suggest the whole process is tightly or completed controlled because the evidence is that this is not so. But there is a very profound and influential form of guidance nonetheless, because the whole process is constrained by the rules of physics, engineering, chemistry, sociology and so on. Genetic variation, reproduction and natural selection have not, could not, and will never produce a creature that has a negative gravitational mass, because negative gravitational mass is not available. The patterns of physics won't allow it. Similarly, biological processes will never produce a social animal that cannot recognize other members of its social group, because that would be a contradiction. It is as if the evolutionary process is repeatedly rolling a dice, so that at first sight one may think the outcomes are random, but then one notices that the only numbers that ever come up are integers, and furthermore they are always in the range 1 to 6. The dice produces random outcomes, but the outcomes are selected from a preconceived 'menu': the numbers written on the faces of the dice. Similarly, evolution can only cause there to come about structure and organisms which embody truths about the areas of discourse in which those structures and organisms operate (areas such as physics, chemistry, engineering, psychology and so on). So there is a guide to evolution after all, a guide in the very way things can be. The question we have to decide, then, is the relative weight of these two elements—the element of uncontrolled variation, and the element of constrained possibility—in the whole story. What is the correct, or the balanced, 'big picture'? Is it lurching, meaning-less change leading to arbitrary outcomes, or is it the emergence of increasingly rich realizations of increasingly deep truths? And if the latter, then was that emergence itself an outcome that was guaranteed?

People presenting this area of science to the general public are often at pains to say that evolution is a physical process which is the outcome of its causes, and as such it has no goal. It is not a goal-directed process; it does not have any aim, such as the aim of producing complex animals. Of course no inanimate process can have any goal in the sense of something consciously aimed at by the process itself, because such a process has no mind. In this sense the claim that the process has no goal is a truism. But it does not follow that such a process is not serving any purpose. It is up to us to employ our purpose-recognition apparatus and discern what purposes it may possibly be serving, or what has gradually been achieved. When we try to discern what it is that has been achieved, it is our own minds that are at work, deciding what is the significance of the various parts of the overall outcome. That outcome includes the opportunity to live that has been afforded to a myriad of organisms, with all the joys and pains thereof, and it includes the ongoing evolution of organisms of every type. One shouldn't think that the new organisms are all more complex than the ones that went before; that is not true. But it is true that the process is able to realize organisms of increasing complexity as it unfolds over time. More and more types of organism emerge and develop, including more and more complex and adaptable types, as well as more varied forms of parasitism and symbiosis and so on. New types of relating also emerge, such as the relation of whole organisms to other whole organisms whom they consciously recognize and respect.

Some scientists want to declare that the main story, or most significant observation, is merely that genes have proliferated. That is not a scientific statement, however. It is a refusal to face facts. It is a refusal to take an interest. In fact, the move from genes only in single-celled organisms to genes in all sorts of organisms, including some of tremendous complexity, is replete with interest and significance beyond mere gene-counting.

People are also at pains to assure us that human beings are not any sort of triumph or end-point or pinnacle. This is welcome as part of a general aim to resist the notion that the world revolves around us. But there is a sense in which humans are an end-point, because our own future evolution will depend not primarily on processes outside our control, but on human choices. This is a significant development, which does introduce a qualitative change to the sequence of life on Earth. The rest of the biosphere will also be very strongly influenced by human choices. It is a great mistake to refuse to recognize and grapple with the unique power and complexity of what humans can do compared to other animals.

Furthermore, the type of consciousness that humans possess is, by any reasonable and objective assessment, a tremendously significant development. It not only underpins the richness of human culture, but has also transformed the very face of the planet, and threatens to be, or possibly is, the cause of one of the fastest and largest extinctions in the history of life on Earth. Its significance is illustrated by the libraries full of literature that you can find across the world; and by the lights you can see from space; and by quantum computing and vaccination and gravitational wave detectors; and by the choir of Trinity College Cambridge singing works of Eriks Esenvalds; and by the Marshall Plan and the Green Revolution; and also by genocide; and by global warming, deforestation, ocean desertification and the ongoing crisis in insect and other populations.

When we look around us at the natural world, complete with the Darwinian mechanism in our intellectual toolkit, we do not need to abandon our sense that living things play their part in something greater than themselves. The purpose of the polar bear, or the worm or the daisy or the mosquito or the humpback whale, is to do with all their capacities and the way they interact with other things at every level of their existence. And it is not true to say that this notion of purpose is either empty or impossible for us to guess at. The notion of purpose is empty only as a contributor to the scientific study of impersonal causes and mechanisms. But such study is neither the only valid form of intellectual discourse nor the only way we live. We live also by poetry and by opening our hearts and minds to the notion that the world around us is full of meaning. And we have precious intellectual apparatus that helps us to discern that meaning, and celebrate the perseverance of the polar bear and the humility of the earthworm, the simple brightness of the nodding daisy, the danger of the mosquito, the majesty of the whale. What beautiful purposes are able to be worked out among them; what ugliness is happening when those purposes are thwarted; what misshapen thinking it is to force mere replication and genetic material into the role of purpose.

CHAPTER 13

THE MAGICIAN'S BOX

In this chapter I shall comment on the modern-day cultural phenomenon of resistance to mainstream evolutionary biology in large parts of the population, especially in conservative religious circles. One of the aims of this book is to offer ways to reduce that resistance, but to tackle it one must face the fact that, to many commentators on both sides of the divide, the divide is welcome, in a sneaky sort of way, because it offers a convenient way to separate science from religion and to denounce the ways of thinking on show in the other camp. In this situation one of the elements of plain fact which will be resisted is the notion that the true picture is less simple. The true picture, the fact about what large numbers of scientists, theologians and everyday people really think, is not a case of two readily demarcated and opposed camps. But perhaps some readers may doubt this, or find it surprising. Let's consider why it may come as a surprise.

A standard part of the tool kit of the traditional stage magician or conjurer is a box with a diagonal mirror inside it. When the conjurer shows the interior of this box to the audience, the box appears to be empty because the part that can be seen from the direction of the audience is indeed empty, and the mirror acts to 'double' this empty part, giving the illusion that it forms the whole.

Liberating Science. Andrew Steane, Oxford University Press. © Andrew Steane (2023).
DOI: 10.1093/oso/9780198878551.003.0013

I think that some readers of this book may be under the illusion that modern-day Christian[1] teaching does not accept mainstream evolutionary biology, and offers instead bad arguments about the processes which led to eyes, molecular motors and other structures. This is an illusion. The reality is that modern-day Christian commitment, across the world, and in its various forms, is more comfortable with evolutionary biology than the illusion suggests. Broadly speaking, the picture is the following. Most academic work coming from people with Christian commitments (including biologists, philosophers, historians, linguists, theologians, etc.) fully appreciates that science offers an important set of insights that ought to be welcomed and allowed to play its part in a developing understanding of the big picture of the world and human life in it. A mainstream line in theological effort here judges that the Darwinian or neo-Darwinian story is, on a deep level, a story in which the biosphere developed depth and richness by the operation of its own innate nature, and this is a remarkable truth and unthreatening to theism, rightly conceived. God is not inferred from a set of miraculous events in prehistory, but rather trusted to furnish or guarantee an ultimate meaningfulness of this rich, astonishing, painful, delightful, epic, persevering exploratory process. Meanwhile there are also significant parts of the worldwide Christian church which adopt a defensive, unintelligent and sometimes ungenerous approach (ungenerous because slow to admit the honesty and quality of the intellectual work of others) and come to other conclusions.

---

[1] I have picked *Christian* here because this is what I am competent to comment on. The label is itself contested because it is used in different contexts to mean very different things; I hope the reader will be receptive to the evidence that there is this breadth, with both good and bad growing side by side in the same field.

Ordinary members of Christian churches mostly do not think it very important exactly how life on Earth developed, but in so far as they have views about it they are willing to be directed by their general education and by their leadership. The leadership in the Roman Catholic, Eastern Orthodox and many of the Protestant traditions is welcoming to Darwinian insights and the membership has included a good number of biologists doing sound work in this area. The Pentecostal movement is varied, and the Evangelical movement is chequered. Concerning the Pentecostal movement, 'varied' is the right adjective because this movement (very large worldwide) is characterized by a fluid and flexible receptivity to local culture in each place, and the local concerns are often to do with ethnic minority groups and marginalized people seeking mutual encouragement. They are not primarily concerned with scientific matters and in consequence the leadership tends to take a line of least resistance on scientific questions. In practice that does produce, I think, an assumption that a literal reading ought to be imposed on the text of the Bible, and the movement will have to learn to do better than this. However, there is sufficient, and sufficiently weighty, scholarship and sheer wisdom on show in other versions of Christian commitment worldwide as to allow mistakes of this kind to be corrected. The conservative end of the Evangelical wing, meanwhile, continues to resist, and feel threatened by, some of the mainstream science. But large parts of the modern-day Christian community do not offer bad arguments about the development of the eye because— well, it's obvious isn't it?—because we learned at school how evolution works.

And yet there is this significant group, associated with terms such as 'creationism' and 'answers in Genesis', which is being thoroughly misled by its leaders and its websites. I am not an expert in this phenomenon and its causes. I don't know whether it is best described as tragedy or comedy or just plain old heresy. The latter word, whose meaning is 'error', may help us here because it may help the people sitting in those congregations and gathered around those websites to know that they have been caught up in a distorted approach to Christian discipleship and to the Bible. They are largely mistaken concerning the literary skill on show in the Bible as it employs a wide range of genres in order to communicate.

Now it so happens that the opening chapter of Genesis does present a process of development and in this sense a broadly evolutionary creation. There is increasing complexity and a world which itself 'puts forth' and supports the living things which are a part of it. But the literary genre of this account is not literal and direct, nor could it be. It communicates by poetic vision and contemplates the role of the elements of the natural order as the brush-strokes of creation sweep across time. Equally, the story of the Garden of Eden tells of a first human made from the dust of the Earth, which is exactly what we are made from, but it does not follow that this has to be seen as a sudden event. It is rather an insight into human nature, in which the timescale is broadly irrelevant. The process is miraculous whether it took fourteen billion years or fourteen microseconds. Equally, this story cannot stand in contradiction to the account of the six days, since its genre is not like that: the garden is not a literal garden but a way of contemplating human nature and the human predicament. Eden's 'Tree of Life' is not made of hydrocarbons.[2]

---

[2] This footnote is for the benefit of readers wishing to understand the role of Adam in the genealogies listed in later parts of the Bible. As I understand it, those compiling

The opening six-day narrative of Genesis is a way of seeing the world as ordered and developing without being completely tied down. It is a world whose conditions make it possible for meaningful, lively existence to take place. The account is in contrast with other ancient stories of struggle, chaos and fights among super-beings. It conveys the deep insight that the physical world is a world in its own right, not a veil for gods and monsters. And there is a striking derived creativity as the plants grow and the animals teem. The phrase 'Go forth and multiply' is an invitation to the creatures to live in their own right, not controlled from above, exploring the biological niches and filling them by reproduction. In other words, it is a statement from an ancient Hebrew community which saw the natural world for what it is, and read it broadly correctly. That community did not have the accumulated knowledge now available to us, which shows that many species have died out and others descended from them have come to live in their turn, but this more complete knowledge of the physical mechanism does not, and cannot, contradict the summary of what has been achieved through that mechanism.

As I say, I do not know the details of how it came about that large congregations misread and mishandle this part of the Bible, and teach their children to be suspicious of science. But it is clear

these genealogies did not have the kind of mindset which modern people take for granted, where the distinction between history and story is held to be a sharp distinction no matter what the context. Whereas we tend to emphasize the scientific approach to understanding pre-history (i.e. the period before written records), biblical authors adopted an approach where the important truths are told through stories. Their reference to Noah and Adam in a genealogy of a historical person such as David or Jesus is a way of connecting people back to the prehistorical past from which we all descend. It is a simple way of saying, of any given person, that they share our common humanity. When writing that way, the compilers of the genealogy did not need, and therefore did not form, any strict distinction between the stories developed in oral tradition to think about the origins of human life in general, and the record of specific individuals in the documented past. They cheerfully merged the two.

that one element has been a kind of intellectual goading coming from some scientific authorities, in which science is presented as a package-deal with naive and dismissive treatment of religious language, rather than science as part of an overall attitude of using our intellectual gifts in the service of good, which includes willingness to handle the Bible and all our inheritance with wisdom and discernment.

Thus, as I have acknowledged, there is a struggle going on in this area. But that struggle is not characteristic of the broad sweep of Christian thinking. The broad sweep is happy to accept that life developed slowly and gradually in all its varieties, by the trial-and-error process we know as Darwinian evolution. Indeed, the worldwide Christian community has a strong history of serious and high-calibre intellectual work, including large amounts of financial and infrastructure support for scholarship and education as well as encouragement to individuals to pursue intellectual endeavour, and this has included work to a high standard in all areas of science. This is a complex history and I don't intend to deny that it has also included injustice and blindness, but the considerable achievements should not be ignored. These include the legal work which was done in the period 100–1000 CE to widen the recognition of the moral worth of individual people of all kinds, and to counter superstitious views about natural phenomena in the population at large. The Christian community also built up the concept of a university and founded and nurtured many of the greatest examples. (Take a look at the names of colleges in England's Oxford University and Cambridge University, for example.) It also contributed large parts of the breakthrough into the modern empirical scientific method in the sixteenth century and its energetic application since then. The notion that the Christian community is scientifically illiterate is utterly illusory.

So how did the illusion come about? It came about partly because the struggle in the USA has been difficult and long drawn out, for complex social reasons involving fear and theological shallowness on all sides. Fights capture the attention of onlookers, so it looks as if this is the main story when really it is not. This explains, to some extent, the illusion of universal conflict. But the illusion is also owing to a conjuring act in which any public speaker who so chooses can show to the world at large the intellectually empty side of religion, and keep the intellectually full side hidden from view. By 'intellectually full' here I mean both the contribution to science of all kinds, including evolutionary biology, from people with well-reasoned religious commitments, and also the long and careful scholarly tradition of discussing what religion is and what religious language and categories can mean, and its practical outcome. Readers who have been misdirected by the magician's box will, I suppose, wonder what this practical outcome is. The practical outcome has included some of the largest literacy programmes in human history, support for scientific research, the promotion of free early-years education for all, reform of working practices, sport, art and universal healthcare—things like that. But the conjuring act of hiding all this behind a mirror has been carried forward with great persistence, until it has achieved the dubious 'success' of hoodwinking the general public. My question to such speakers is: why? Why do you want to hoodwink the general public?

In view of this strange and unsettling state of affairs, I felt I had to add this brief chapter in order to make it clear that I, and everything in this book, and most theologians for that matter, have no problem in receiving the neo-Darwinian evolutionary account of the development of life on Earth by variation and natural selection, as it is laid out and developed in studies and publications by the mainstream scientific community. But what we like is straight evolutionary biology: the thing itself, not illogical claims about what it all means.

CHAPTER 14

STEPPING OUT

S o what is the correct 'take' on evolutionary biology? No one
can claim to have the whole truth on a question like that,
because it is a question which overlaps with questions about the
role and value of life itself. But we can at least agree some of the
more straightforward elements of a correct interpretation. One
such element is that the process whereby the natural world has
developed is just that: a process. So what? Trees grow gradually;
so did life on Earth. The process was neither meaningless nor
largely random. So let's accept its creativity and let's step into
the freedom we have been given through the outcomes of this
process. I mean the freedom to enjoy all our human capacities,
no matter how they came about.

In the following my aim is to give to the reader a basis on
which to step out into what freedom we have. We are not totally
free, but we are not completely controlled puppets either, and
there is a vast territory which we can explore. I want to show
why the willingness to recognize the connection between *being*
and *love*—the connection which lies at the heart of religious
response, when it is good religion as opposed to bad religion—is
not unsettled or reduced by what has been discovered so far about
the process whereby life developed. I am writing for the reader
who is interested in evolution as a fascinating process for its own

Liberating Science. Andrew Steane, Oxford University Press. © Andrew Steane (2023).
DOI: 10.1093/oso/9780198878551.003.0014

sake, and for the reader who doesn't feel strongly about this but has absorbed some message that 'science displaces religion' or feels inclined to say (as I heard someone say) 'I don't believe in God because I believe in evolution'. I want to show how that idea is not just wrong but ridiculously off-base. It is equivalent, in its logic, to saying 'I don't believe in mathematics because I believe in arithmetic'. It really is as bizarre as that. What such statements signal is the success of public misinformation campaigns about the meaning of religious vocabulary. Thoughtful religion is concerned with the reasonable intuition that qualities such as kindness, justice, courage, wisdom, mercy, hope and fun are not arbitrary offshoots of natural processes, nor human inventions, but rather are fundamental to the very way things are, to existence itself.

First I will deal briefly with the dispute currently going on in American public life, and elsewhere, under the name of 'intelligent design'.

Evolutionary biology is a scientific discipline which continues to develop, and there is much to learn. Scientists are rightly interested in the areas where our knowledge is incomplete, and from time to time someone will make a claim that some particular feature cannot have come about by the neo-Darwinian route.[1] Such a claim is a scientific claim, because it is amenable to scientific assessment. As things stand, no such feature has been found, but the mechanisms involved are sometimes quite subtle.[2] That is,

[1] By 'neo-Darwinian' I mean, broadly speaking, the accumulation of small genetic changes over many generations, amplified in any given population by reproduction when the change enhances reproductive success, and in which the changes are largely uncontrolled. There is evidence that cellular processes can include a mechanism whereby a given cell's environment can influence change in that cell's genome in non-random ways. It is a matter of debate whether the term 'neo-Darwinian' includes this possibility. I think it can, so I am using it that way.

[2] For example, epigenetic changes can persist over sufficiently many generations as to influence what genetic changes subsequently have reproductive success. It is not yet

although there remains room for argument, no one has found a biological organism or organ or metabolic process or cellular component which clearly could not have come about by the accumulation of small changes on the relevant timescale. But it is fair to say that the fossil record remains puzzling. The remarkable increase in diversity in the early Cambrian period, for example, is far from understood. Scientific journals also document many other open questions in evolutionary biology. So there is nothing improper about proposing a hypothesis for a mechanism at work in developmental processes in biology in addition to the standard neo-Darwinian route, if one can also furnish a reasonable case and a means whereby the hypothesis can be tested. However, it is improper to overstate the evidence or misrepresent the main thrust of the evidence, and there is something wrong about presenting arguments to the general public which can be refuted by a modest application of logic. The 'something wrong' is either a failure of intelligence or a failure of intellectual integrity.

An example of the latter (an argument which can be refuted by a modest application of logic) is the argument that a complex organ which requires all of its parts in order to function cannot develop by gradual change unless the parts are themselves individually functional—that is, can be of benefit to the organ each on their own without the others. The argument (which I am about to refute) goes as follows:

> Consider a molecular motor such as the one used by a single-celled organism to drive a rotating part (a flagellum) by which it swims around. This motor combines several complicated proteins all

clear whether such facts have a major or a minor role in large-scale evolution. For further information see, for example, Eva Jablonka and Marion J. Lamb, *Inheritance Systems and the Extended Evolutionary Synthesis* (Cambridge University Press, 2020).

working together. But those particular proteins are not able to help the cell in other ways. So how did they ever come to be there? If they were produced one by one through genetic change over many generations, then each on its own would have made life more difficult for the cell (because it is a useless part on its own) so would diminish reproductive fitness and would appear less often in subsequent generations. So there is no gradual route for them all to appear. They must have come together by some more rapid, orchestrated change.

This argument is associated with the phrase 'irreducible complexity'.

The argument is refuted by observing that parts of the cell, such as proteins, can themselves be modified through gradual change over generations. Suppose for example that items A,B,C are gradually introduced over generations of a cell, each functional in some way, and subsequently a new item D is introduced following some genetic change. It can happen that the role previously played by C can be played by A,B,D together, provided they are all present. In such a case C becomes an unnecessary ingredient so subsequent generations will manage better without it. So there can now arise cells containing the structure A,B,D which requires all its parts in order to function, and one of them (D) may have no other useful role. Now it may be that this new structure A,B,D is especially useful to the cell: maybe it is a molecular motor, or an ion pump or something like that. The items A,B will themselves be subject to occasional change, leading to further generations with slight changes, let's say making E and F. If those changes make the motor work better, then they may well be amplified in the population, even if E and F on their own are of no use to the cell. So the end result is E,F,D, a structure which requires all of its parts in order to function, and where each of the parts on its own is not functional, and the result was arrived at by gradual change over many generations. And furthermore there was no need for forward planning or coincidence: at each step along the way some small change of immediate benefit was involved. Thus a structure which was described as having 'irreducible complexity' can in fact be put together by the standard neo-Darwinian route.

This example shows that complex organs or processes that require all their parts in order to function (parts that would be useless otherwise) can come about by the neo-Darwinian route, so it would be false to make any sweeping assertion to the contrary. This does not render it impossible that some particular example may have come about some other way, but it shifts the burden of proof, and one should not misrepresent the main thrust of the evidence accumulated to date by large amounts of meticulous work in the scientific community.

The main thrust of the evidence is that the neo-Darwinian process is what happened. The implication is that the development of richer and more variegated forms of life on Earth did not sometimes take place by sudden influxes of information, as far as we can tell. However, evolution is sometimes much more rapid than a model based exclusively on random changes would allow. This can be understood as a response to the environment; I will discuss it shortly.

In the previous four paragraphs I dealt with one way in which evolution might be misrepresented. Now I will turn to four other ways. Each amounts to an illogical claim. First, there is the claim that evolution is largely a random sequence with random results. This is so far from true that it is amazing that people ever maintained it, as I will show in a moment. Second, there is the claim that the relation between genome and cell behaviour is a one-way cause–effect relationship. In fact it is not that simple. Third, there is the claim that the mental and social life of humans is largely caused by genetic variation and natural selection. I will argue that whereas those elements have a part to play, the mental and social life of humans is largely the result of other considerations. Fourth, there is the claim that science can tell us whether or not the process of evolution serves some purpose. In the sections that follow I have focused on these four listed items for the sake of clarity. It helps to be clear on what claims we can jettison. As these items fall away, a clearer view of our place in the world then comes to the fore, and it mostly speaks for itself. In any case, by dropping illogical claims we will give to ourselves and our children the chance of a more insightful and a more rounded sense of our place in the world, and consequently a better chance of taking good decisions in public policy as well as in scientific research.

1. Randomness

Let's begin with the role of randomness. One obvious fact about evolution is that simple things came first, and complex things came later, alongside further simple things. So: not random. It had to be like this, because the complex things take up and combine structures that have previously been furnished by the simpler things. In this respect the evolutionary sequence is not random. But the really interesting thing here is not that complex things came later, but that complex things came at all. *That* is a very striking fact, and it was guaranteed to occur.[3]

---

[3] I mean guaranteed to all intents and purposes, like the way most thermodynamic processes are guaranteed by the law of large numbers (the law of large numbers is the observation that processes involving probability have some predictable features when many examples are involved).

Furthermore, the nature of both the simple and the complex things is not random except in secondary respects. Notice, for example, the simple fact that our bones are round, of roughly circular cross-section, mostly not square or triangular, and in this respect they are like the trunks of trees and the stalks of flowers and the antennae of insects and the arteries of mammals and the bodies of worms and countless other examples (Figure 14.1).

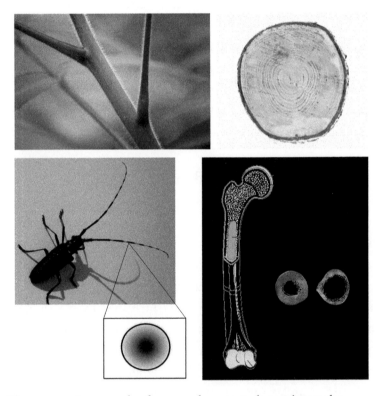

Figure 14.1. An example of non-randomness in living things: the common near-circular cross-section of plant stems, tree-trunks, antennae and bones. But this is just one of a vast array of examples, spreading from electrical signalling and chemical-reaction networks to the structure of whole organs, bodies and social groups.

Is this all some sort of massive coincidence? Just a fluke amongst a vast randomness? Of course not. It is because the cylinder is a naturally strong shape which makes efficient use of materials, so living things with genetic programming for cylindrical shapes tended to prosper better than others and pass on their genetic characteristics to more offspring. To say 'evolution gave us round bones' is true but empty. If we want to understand *why* bones are round, most of the task is to understand the mathematical and engineering properties of the cylindrical shape. It is not enough merely to say that this shape has a better survival value than others—one should seek to understand why it does, or what are the truths embodied in it that give it this property.

So we have that the shape of our bones is not random. There is an element of randomness—skeletal structure varies in size and in detail—but broadly speaking our bones are an expression of certain non-random facts about mathematics, engineering and materials, along with the physics of mass, acceleration and gravity, and the strength of the local gravitational field.

Once you have had a simple example like this drawn to your attention, you can quickly generate a myriad more[4] and thus discover quite how misleading it is to assert that evolution is largely random. Experts on evolutionary biology will acknowledge this once it is brought to their attention. Indeed, there is a large amount of work on evolutionary constraints in the professional scientific literature. The issue concerns how the

---

[4] Nerve cells respect the laws of electromagnetism; heavier animals have larger muscles; the genetic code is neither too simple nor too complex; the shapes of plants satisfy engineering principles for load-bearing structures; heartbeats generate sufficient pressure to circulate blood; fish gotta swim and birds gotta fly; etc. For an exposition of more sophisticated examples, see Conway Morris, *Life's Solution: Inevitable Humans in a Lonely Universe* (Cambridge: Cambridge University Press, 2003).

balance of the subject is presented to the general public. The acclaimed evolutionary biologist Ernst Mayr devoted space to the constraints on evolution in *What Evolution Is* (Basic Books, 2001), indicating his judgement that they are an important part of the whole story. My aim here is simply to help the reader to know it too.

2. Two-way traffic

In this section I will briefly present the fact that contemporary research in biology yields a much richer picture of what is going on than the one commonly assumed by the general public. The process of evolution is significantly misunderstood when it is thought that random point defects in the genome are the only route for change. In fact, cell division includes quite a rich range of capacities for reconfiguring genetic material and putting old tools to new uses. And as I already remarked in Chapter 12, the relation between genetic sequences and proteins is not one-to-one but depends on context: the same sequence may code for different proteins in different circumstances, and different genes or combinations of genes can result in the same structures. But my present point is not merely that biological mechanisms have a certain amount of complexity. The point is that the evidence suggests the relationship between genome and cell is not simply one-way, as if the genome simply tells the cell what to do. It is not like this. To a significant degree, the cell tells the genome what to do. That is, the cell and its environment have a significant influence on what will be the role and the effect of given genetic sequences in that very cell. Also, there is a much greater capacity for rapid adjustment of the genome in response to changing conditions than would be the case if the relation between genome and cell behaviour were only a one-way cause–effect relationship. As Jablonka and Lamb put it, 'There is now good experimental evidence, as well as theoretical reasons, for thinking that the generation of mutations and other types of genetic variation is not a totally unregulated process'.[5] It has been observed that 'bacteria that have lost their flagella through deletion of the relevant DNA sequence can evolve the regulatory networks required to restore flagella and so restore motility in response to a stressful environment in just four days'.[6] Jack *et al.* find that cells 'possess specific mechanisms to optimize their genome in response to the environment'.[7]

I think it is not yet established how great a role this kind of two-way relationship between cell and genome plays in the life of each living thing and in the overall development of life on Earth. But there is a good chance that it is highly significant because it offers an avenue of response to a stressful environment which otherwise might be closed completely, and the difference between passing through a narrow gap to a new territory and dying in the gap can be a great difference. In evolution it is often *adaptability* rather than mere brute strength that is key.

What is certain is that the overall picture is not one of little mechanical machines with programmed responses; it is much more fluid than this, with organism and environment mutually feeding back and moulding each other.[8]

---

[5] Jablonka and Lamb (2005), 80.
[6] Taylor, Mulley, Dills *et al.* (2015).
[7] Jack, Cruz, Hull *et al.* (2015).
[8] For further details and references to the scientific literature the reader is referred to the survey by Iain McGilchrist in chapter 12 of *The Matter With Things*, to which the present subsection is indebted.

3. Causes and moral discernment

Next let's turn to the third illogical claim: the claim that the causes of human experience and behaviour are mainly genetic variation and natural selection. It is the notion of *causation* that I want to get at here. Please let the reader not misunderstand me: I already made it explicitly and repeatedly clear that I am not questioning the main facts of the Darwinian process as they have been discovered by a rich and welcome body of mainstream scientific effort, nor am I proposing other types of process here and there. I am engaged in reflection on what the process is really like. The structures and behaviours of living things have been arrived at, on Earth, via genetic variation and natural selection. But why those particular structures and behaviours? If we can answer that question (and we can, at least in part), then what kind of answer do we get? It's not enough to say merely 'Oh, it's because they made the genes that coded for them more common'. I want to ask, first: what aspects of those structures had to be as they are irrespective of reproductive success, and, second: for those that could have been otherwise, how do they operate and what is their characteristic form or pattern?

'Natural selection' is simply a name for the fact that types of life less suited to reproductive success will (obviously!) reproduce less, and the selection is acting on the inherited material (primarily genes). Natural selection does not cause the properties that enhance reproductive success; it is not a cause at all, it is just a name for a filter-like aspect of the process. Our nature, and the nature of other living things, is to do with what principles of physics, chemistry, engineering, biology, social science, psychology and spirituality are on show in, and embodied by, living things. We are caused by the combined operation of all these principles, which constrained what the products of the neo-Darwinian process could be.

Now I realize that the reader will be wary here. I don't intend to deny that the term 'natural selection' could be used as a summary term for the whole process with all the contributing factors which I just outlined. One can use it for that. My aim is to invite thought about what is really going on. To show you what I have in mind, consider the following example. Suppose a process resulted in there coming to be an organism which has the means to move itself around: it has legs, or fins, or wings. This very thing happened in the gradual evolution from early life forms to plants and animals. So the evolutionary process is the cause of this organism having legs. But the evolutionary process is not the cause of the landscape which the organism discovers as it runs about on its legs. Genes and inheritance did not arrange the location of the water sources and the mountains and valleys; Darwinian evolution did not make rocky cliffs and sandy deserts and arrange their location. Things like that are owing to tectonic movements and weather. The point is that when, as a result of an evolutionary process, an organism has a capacity—in this example the capacity to move about—that process does not and cannot dictate the landscape which that capacity allows the organism to explore. (Life can transform the physical landscape up to a point but this does not deny the main thrust of this paragraph.)

In human life we have capacities such as moral judgement. We employ this capacity to explore the moral landscape. The capacity is an outcome of evolutionary processes. The moral landscape is not.

Now suppose someone were to raise a question such as 'why are the Himalayas high?' and then go on to claim that it is because there is an evolutionary pay-off: only strong birds can fly over them, or something like that. We would not take such a claim seriously; the 'reasoning' is not reasoning at all. No one makes that mistake about the Himalayas, but a similar mistake about more subtle territory is very common. Consider next a capacity that humans have, and that some other animals may have to a much more limited degree, namely the capacity to employ simple logic about number. We can do arithmetic such as 'two plus three equals the same amount as four plus one'. But *why* is two plus three equal to four plus one? Is it because there is an evolutionary pay-off? Is it so that we will live longer, or make better decisions, or get more mates? Or is it an innate property of mathematical logic that no evolutionary pay-off can influence one way or the other? The answer is the latter: it is an innate property of mathematical logic that no evolutionary pay-off can influence one way or the other.

Now it is true that being able to reason about number is a useful ability; anyone who can do it is likely to fare better. Making a mistake might get you killed by a predator you didn't account for, or might make you fail to provide enough food for your offspring. That is all true, and it is about the capacity to understand number and reason accordingly. But it is not about the laws of arithmetic themselves. They are not caused by evolution; they are a kind of landscape which is what it is; all we can do is navigate that landscape if we can.

The idea that arithmetic rules have been created by a process of natural selection is not, as far as I know, being suggested by anyone.[9] But now we come to it. Many other parts of human culture have been said to be created by a process of natural selection. Things like ethics and morality are said to be subtly coded assumptions and behaviour whose ultimate cause is that they result in more replication, on average, of the genes which tend to make brains approve the moral rules. Indeed, a vast literature has grown up around this notion. It is a notion that has the same logic as the claim that the Himalayas are high because it results in stronger birds in the long run. Of course the human social example is a lot more complex, and I admit that in this case there are genetic advantage factors in the mix, but I am making the considered and rational claim that they are a small part of the mix. The fact that murder is wrong is a brute fact; it is part of the moral landscape which no gene can do anything about; what we have is a precious capacity to perceive that moral landscape and a responsibility to navigate it. Let's put to one side the special cases; I am not talking about end-of-life decisions for those whose life is already curtailed, nor defensive action in extreme situations. I am talking about ordinary anger and hatred and violent crime. *Why* is murder wrong? Is it the case that murder is really neither right nor wrong and our moral insight is no insight at all, but just a coded passion, an offshoot of other thoughts and passions, one we have because ancestors who did not consider it to be wrong had fewer surviving children? The answer to this question is not to be found in any study of reproductive success, just as reproductive success cannot tell us whether a mathematical proposition such as the Riemann hypothesis is true. The morality or amorality of murder

---

[9] Well I guess some people may have tried it, but I doubt if they arrived at a satisfactory way to say that the result of adding two pebbles to three pebbles could possibly be seven and a half pebbles, or a hundred pebbles, or two ice-cream cones and a kangaroo.

has *strictly nothing* to do with biology and biology will never tell us anything about this distinction one way or the other. Biology can help us understand our propensity to murder, or not to murder, but not whether the propensity is good or bad, right or wrong or neither. But it is reasonable to hold that we do have moral insight, as we have also mathematical and other forms of insight. And it is well established that murder is wrong because it is a failure to recognize and respond adequately to the value that is embodied in another, and a failure to consider the wider ramifications; it is a failure to rise to the stature of what being a person involves. There are further such reasons and considerations; they are the ones discussed by works of literature and history and moral philosophy. *Evolutionary biology has strictly nothing to say on this question.* That does not mean it has nothing to say about human behaviour. But what it shows us is the building process whereby humans acquired the capacity to move around on legs and find the Himalayas towering over them; what it shows us is the building process where humans acquired the capacity to have moral sensitivity and find the prohibition of murder towering over them. And biology also rightly warns us that none of our capacities are going to be perfect, and that includes our moral perception.

I hope the reader will now see that large amounts of writing and cultural fashion of the last thirty years, setting out from the assumption that evolutionary pay-off is the (unconscious) rationale for beliefs formed by human brains, is unscientific and incorrectly reasoned. The reason we believe the Himalayas to be high is not some unconscious desire that they should be high, nor fear that they might not, but that we have the ability to perceive that they are high. Similarly, the reason that we form a huge number of important conclusions about how life should be lived is not that we are unconsciously programmed to follow some other agenda to do with genes, but because those genes have facilitated the building of bodies that possess logical, aesthetic and moral perception—bodies that can perceive logical, aesthetic and moral landscapes. We perceive and inhabit those landscapes in all their beauty, subtlety and difficulty.

The processes of genetic reconfiguration and mutation, followed by amplification and reproduction, act like a kind of search algorithm, searching for solutions to the problems of living. The nature of the solutions—that is, of the living things themselves—combines on one hand the property that they had to be arrivable at by such a process, and on the other the property that they had to represent viable solutions. And that means they had to present truths not untruths, such as the truth that carbon chemistry is richer than helium chemistry, and the truth that osmosis occurs across semi-permeable membranes, and the truth that light travels in straight lines, and the truth that mutual trust gives better prospects to a social group than mutual suspicion, and so on. These high-level truths, along with the low-level molecular dynamics of DNA, together constitute the causes of the structures of living things.

How often do we hear the phrase 'evolution gave us ...' or 'owing to evolution, we tend to ...'? It is the phrase people use as a prelude to some statement about our thinking and behaviour, under the illusion that this way of talking about human nature is scientifically grown-up. I have heard it so used by undergraduate students at Oxford University, for example, which illustrates how the way of thinking has permeated culture and infects intelligent people along with everyone else.[10] But this attitude is usually a prelude to not thinking, to not trying to understand the truths that have shaped us. It often introduces a way of thinking which assumes that the propensity to pass on genes will give us the primary clue to every facet of our behaviour.

To repeat: the latter assumption is wrong. Consider again the mathematical case. The correctness or otherwise of a mathematical assertion has got strictly nothing to do with the propensity of anyone's behaviour to cause genes to be more abundant in the future. Our tendency to think that 'nine added to sixteen makes twenty-five' is not a mere genetic accident; it is the result of a much more interesting and more subtle combination of facts, with genetics at one 'end', so to speak, and the mathematical landscape—a landscape of ideas—at the other. We need to see that similar statements can be made about much of human culture and human nature. Our abilities to reason, and to create art, and to form value judgements are influenced by our genetics, but the correctness of any given piece of reasoning, and the quality of any given artwork, and the soundness of any given value judgement all have nothing whatsoever to do with genes or anything else in biology. Thus large amounts of human culture have got *strictly nothing* to do with any propensity to cause genes to be more abundant in the future.

---

[10] Here are two further examples. (i) An academic referee reported: 'I come across students telling me that ethics and value is just the trace in our minds left by the evolutionary process all the time'. (ii) Chatting on a car journey with a teenage friend of my daughter, she said she had heard that human ideas about morality and value are simply assumptions or instincts caused by natural selection; when I put it to her that what we had inherited is not an arbitrary set of opinions but rather a capacity to perceive value, she warmed to this idea and said she had not heard it before.

The idea that we have to turn to evolution to explain complex human behaviour sounds quite scientific, and of course it is not completely wrong, but it has for some decades now given rise to a sort of parascientific literature whose proponents do not see how painfully illogical their way of proceeding is. Marilynne Robinson, in *Absence of Mind*, points out examples. Here is one:

> Family feelings are designed to help our genes replicate themselves, but we cannot see or smell genes …. [O]ur emotions about kin use a kind of inverse genetics to guess which of the organisms we interact with are likely to share our genes (for example, if someone appears to have the same parents as you do, treat the person as if their genetic well-being overlaps with yours).[11]

So writes Steven Pinker, and one should acknowledge that he is aware that the line from the genes to the behaviour is quite subtle. But let's see what the truth of family feelings really is. Take a parent and child for example. What happens is that the parent experiences a sense of delight in witnessing one of the most wonderful marvels of the universe: the growth of a person. The parent also experiences all the usual difficulties of interpersonal relationships. Why does the parent experience all this? It is because the parent has a certain capacity: *the capacity to recognize a person, a subject, one who has thoughts and interests of their own, with all the value that this entails. That* is what their genes and their experiences have conferred on them: an ability to deal in truth, not an ability to detect genes hiding under the bonnet of an automaton. It is an ability to deal in truth with all its rewards and difficulties. This, not the notion that we are chemical robots, is the scientific conclusion

---

[11] Steven Pinker, *How the Mind Works*, 324–7, 456–9.

here; the biological conclusion, the correct analysis of the result of evolution.

This is not to deny that we are also steered by less subtle issues, and 'Family feelings are designed to help our genes replicate themselves' may be OK as a reference to certain instincts that do play a role here. The problem is that it is widely invoked as part of an illogical programme: the programme to deny that 'one plus one' is actually equal to 'two' in favour of saying that such a thought is no more than a coded way to get my genes out there into the pool. Of course Pinker would not pick that example. He picks the example of the parent–child relationship. It is not 'I plus thou equals love', he says, but a 'a kind of inverse genetics to guess which of the organisms we interact with are likely to share our genes'. Perhaps he writes that way because he is not willing to treat seriously the formula 'I plus thou' as one which deals in objective reality, or else he is aware of it but finds it to be peripheral to the story he wants to tell. But this is an objective reality, I claim on the basis of evidence and reason, and the fact that a parent's biological children share half the parent's genes is of only modest relevance (as witness the phenomenon of abusive parents on one hand, and adopting parents on the other). Also, 'I plus thou' may not be marginalized in any account that aspires to objectivity, or that merits to be communicated from one person to another, because to marginalize it is to adopt a bizarre, unscientific and lop-sided refusal to engage with what is actually going on. What is actually going on is that humans, for all their imperfections, have real capacities to think, give attention to each other, and recognize the absolute value of one another.

The large and thoughtful wisdom tradition to be found worldwide will agree that these more rounded kinds of attention that we pay to the physical world and to each other are indeed central and trustworthy, not peripheral or illusory. I am presently engaged in pointing out that the attempt to deny this in the name of rationality is itself a failure to reason correctly. We are indeed the product of our physical history, but that physical history has resulted in our having certain capacities, and we do indeed have those capacities! The capacity to have empathy, discernment, self-sacrifice, joy, mindful awareness, the ability to recognize each other as people, and so on. So let's rise to the task of employing these capacities appropriately.

The need to do this is especially acute in contemporary western culture, as David Bohm, Marilynne Robinson, Iain McGilchrist and others have argued. All their work should encourage us to step out into the freedom which a more balanced perception offers— the freedom which I am urging in the present chapter.

To conclude: our biology furnishes us with various abilities to explore the physical world and to explore the worlds of ideas and of art and of moral value and of interpersonal relationship, but it does not create the truths of physicality and of logic and of art and moral value and interpersonal relationship. The identity of each of us is very much influenced by our exploration of those truths, those worlds. So it is far from true that we are mere products of biology. We are not the products of genes merely; we are the products of all the types of influence which bear on us. Products of the principles embedded in the natural world at every level to which we have access. Children of dust, yes, but children also of love.

And—marvel of marvels—we are ourselves a channel of that love to the next generation.

4. Purpose

Now for the fourth illogical claim: the claim that science can tell us whether or not there is a purpose to the story of life on Earth. Chapter 12 already touched on aspects of this. In that chapter I argued that the claim that gene-propagation is the main role or purpose of living things does not logically follow from scientific study, nor is it suggested by scientific study. The claim rests on an unbalanced way of handling the network of scientific areas which have insights to offer concerning processes in the living world. What I want to do now is discuss more generally how scientific study interacts with questions about purpose.

Suppose someone were to form the judgement that a physical process such as Darwinian evolution serves no purpose. Some people do form precisely that judgement. Roughly speaking, they want to say, 'it is just a more-or-less hotchpotch sequence of events, serving no particular purpose; it is just what happened, blindly making life forms as the wind makes waves. No aim is being realized'. It is not illogical to say this—to have this opinion. But it would be illogical to claim that this is a scientific or an objective judgement. Such a claim may be someone's opinion, but it is not a scientific claim and should not be presented as if it were. It is not a scientific claim because science does not operate in those terms. Science is a way of finding things out, and it is an intellectual discipline, but it is not a discipline which can provide a judgement about the purpose of a whole; it can only elucidate the role played by the parts in making up that whole. By 'a whole' I mean, for example, the whole process of evolution.

Hence the opinion that there is no purpose to evolution is not a scientific opinion. Here by 'not scientific' I don't mean in the sense of contradicting science or that a scientist couldn't hold it, but in the sense that science cannot itself furnish the opinion, because science does not furnish methods to test it. The statement 'there is no purpose' is not a scientific but a religious statement. That doesn't mean you are not allowed to say it. You can say it if you like. But you should not kid yourself that you are being objective. Assessments of this kind are subjective.

The idea of purposelessness has been explored in the arts and in the philosophy of absurdism. Such a philosophy seems to want to show signs of courage and honesty, but it is hard to argue how it could consistently prefer courage over cowardice or honesty over duplicity. It seems to struggle to adhere to rationality, too. If so, then it cannot correctly be called scientific or an adequate champion of science. It tends to result in statements that are demonstrably nonsensical, such as the idea that random changes in genes in the past have dictated the nature of humans now. That is like saying that the flowing metal in an ironworks dictates the shape of the mould into which it flows. Or it is like saying that a pair of legs dictates the shape of the terrain an explorer encounters.

None of the above renders the 'purposeless' conclusion demonstrably untrue. Perhaps it is correct—maybe there is no purpose. But such an opinion is going to have a hard time being consistent with the ongoing practice of science. To announce that processes in the natural world serve no purpose is tantamount to saying that our attempts to understand them also serve no purpose, because, after all, we ourselves are elements of the natural world. But if there is really no purpose to doing science, then we may as well stop doing it.

So then maybe someone will say that there is a purpose to our scientific and other activities: they have whatever purposes we give them. In this way one ends up with a vision in which human will is at the centre. It is a vision which can function, but I consider it to be an ugly vision in comparison with the alternative. The alternative is that there is a purpose to the world and ourselves in it, one not invented by us, one which we can decide to try to discover and voluntarily promote with our labours.

CHAPTER 15

ANGELS WITH DIRTY FACES

One of the ways in which any culture or organization can go wrong is when it exerts, whether deliberately or not, an oppressive atmosphere, in which people are not free to have their own thoughts and receive those aspects of goodness or truth or beauty to which they are sensitive, and to question assumptions around them which may be doubtful. In the worst cases the situation becomes cultish. It is profoundly sad to contemplate the amount of stress and unhappiness that has resulted as people struggle on, or slowly free themselves from such organizations and such sets of assumptions. Think, for example, of the gentlemen's clubs of early twentieth-century Europe and America which, while innocent in many respects, perpetuated misogyny and privilege and thus not only slowed the expansion of voting rights for others, but also made miserable the home lives of the very members of those clubs, by thwarting the growth of their hearts and souls, their understanding, their aspiration to becoming a partaker of humanity at its best. Or think of the way religious practices, which may have started off as an honest attempt at goodness and beauty, can go horribly wrong as they descend into superstition and control.

And a controlling mindset can be quite subtle. It shapes the type of discourse which people engage in, the type of jokes they

Liberating Science. Andrew Steane, Oxford University Press. © Andrew Steane (2023).
DOI: 10.1093/oso/9780198878551.003.0015

make, the little signals in body language that accompany every conversation, and the way people read the trajectory of human history. It is all about exchanging signals which assure us that we are in the 'in' group, the wise ones, the ones who have not been hoodwinked. One such modern-day phenomenon is the culture of those who 'know' that physical science tells us the origin of the cosmos, and that the Darwinian process is largely random, and that the process of our life is a kind of machine whose motions are either preordained or else merely random, and there is no other possibility. My heart mourns for people caught up in this mindset, especially because many of them have been led to think that all this is 'what science says', and therefore it is forced upon us: to accept it is the only way we can find maturity or join in with the scientific enterprise.

It is important to note the forcefulness here. The forcefulness is not the good kind that comes from a valid sequence of reasoning. There is no valid sequence; the conclusion is illogical.[1] The forcefulness lies in the claim that the opinion being put over is not an opinion but a scientific conclusion. It is that kind of claim to which I object. An oppressive forcefulness was noted by Francis Spufford in a different, but related, area. As he puts it in his eloquent book, *Unapologetic*, parents find with dismay that their children grow up in a culture which denies them space to think because the latest prejudices are 'shouting in their ears'.

---

[1] The claim that the patterns of the natural world (I mean the ones that actually hold, not the ones in scientific models developed to date) can exhibit only determinism and randomness is an opinion, a guess, not a deduction. It is a wild extrapolation from the relatively small amount we understand about living things. When it is said to be not an opinion but a scientific conclusion then we have a fallacy—an example of the *Babel fallacy* (see Steane, *Science and Humanity*, Oxford University Press, 2018).

In the 1938 Warner Brothers film *Angels with Dirty Faces* [2] we follow events in the lives of characters Rocky Sullivan and Jerry Connolly. As boys they had together attempted to rob a railroad car. Chased by police, Jerry is a little faster over a fence and escapes, while his friend Rocky is caught. Fifteen years later, both have turned into brave men. Rocky has the type of courage that a gangster may have; Jerry has the kind of courage that a community worker and teacher may have. When Rocky returns to the neighbourhood after having spent some years in prison following a more serious crime, Jerry finds that the youths whom he is trying to mentor have formed a sort of idol-worship towards Rocky with his tough reputation. He realizes that they risk falling into the destructive way of life that Rocky has followed. Rocky is no arbitrary criminal; he respects his friend Jerry and acts to protect him from a murderous plot. But he himself gets into a shoot-out and kills a police officer. The film reaches its climax as Rocky faces the death penalty. Jerry visits him in prison and asks him to do one final act of courage, on behalf of the young people in Jerry's care. But now it is courage of a different kind. Jerry asks Rocky to put on a show of cowardice before the electric chair, so that the young people who have idolized him will not think so highly of his gangsterism. In this way he will enable them to break away from that way of life. The act will also reveal his own deeper truth. The truth that Rocky more deeply stands for is not the gangster way of life he has followed up till then, but that the young people around him should have a chance to find a more constructive path, and not be bamboozled by the glamour of his reputation as a hard man.

[2] Directed by Michael Curtiz, screenplay by John Wexley and Warren Duff based on a story by Rowland Brown.

In academic life it sometimes happens that a person will write a forcefully argued book which generates a lot of interest and sales, and then in later life realize that the argument of the book was flawed. They might themselves admit this flaw, but it may be that they take no great trouble to see to it that their flawed work is then recognized as such. It remains in the bookshops where it will no longer mislead professional academics but it will mislead adolescents. An example of this is the book *Language, Truth and Logic* by the philosopher Alfred Jules Ayer, published in 1936. The book sets out plenty of sound reasoning but it also includes a sufficient amount of unsound material as to be deeply flawed overall.

The situation is similar for books such as *The Selfish Gene* and *The Blind Watchmaker*. They contain plenty of sound reasoning, but also a sufficient amount of unsound premise and failure to see relevant issues as to be deeply flawed overall. The unsound premise is that one aspect of a physical mechanism of causation tells the whole or the primary truth of what is caused; the omitted relevant issue is that evolution is guided by ontology, and the space of possible viable phenotypes is far from random. Furthermore this space is explored more and more fully and richly as time goes on, so that there is a sense of direction to evolution too. I would like to invite any supporter of the argument of the above-mentioned books to consider the case I have made for these points, and, if they find that it is a well-reasoned case, then to consider what we should be saying to our schools and wider culture about the process of evolutionary biology, because as things stand many people have learned a view of it which leaves them with an inaccurate and seriously misleading assessment of their and everybody else's situation and potential.

CHAPTER 16

SCIENCE AND SENSIBILITY

There was a dancer who,
approaching a stream, stepped
in marvellous speed and balance
upon the many lily pads,
and soon crossed over.

His pupils tried to follow
but floundered in the water.

So some replaced the lilies
with stepping-stones, the better
to help their fellow followers.

But the aim was never
to get across the stream.
It was to become a dancer.

The main theme of this book is science and the public presentation of science. In order to comment further on that in
the next few chapters, it will be necessary first to discuss science
and religious sensibility. This is the subject of the present chapter.
Note, I do not say 'science and religion' because I don't propose to
embark on that larger topic. My aim is to help the reader develop
a correct feel for what kind of thing a religious sensibility is, and
how it operates alongside, or in choreography with, scientific
intelligence. Why isn't it the same as scientific thinking, and if

Liberating Science. Andrew Steane, Oxford University Press. © Andrew Steane (2023).
DOI: 10.1093/oso/9780198878551.003.0016

it isn't the same, why is that not a defect? I shall occasionally use the word *God*, but as I have already suggested in earlier chapters, one should not immediately assume that one knows what the word means. It should be seen not as a label but as a pointer.

In science, model-making is central. It is what science is all about. But in relations between persons, the model-making kind of analysis is not how knowledge grows. This distinction is a helpful way in to understanding the difference between scientific and other ways of learning, and to appreciating that both have valuable roles to play. In scientific model-making we do not engage directly with the world in all its complexity, but instead present to ourselves a simplified version which plays out in a kind of internal theatre of the human mind. In the internal theatre, things are precisely defined and comprehended by us, because we are the ones creating and inspecting the players in the theatre—the physical model with its rules. The model is not entirely a fiction because it is developed to give insight into empirical observations, and to predict new ones, and in this sense it is a conveyor of truth, or of aspects of what is true, or a helper towards truth. But even so it may entirely fail to grapple with central aspects of the things being modelled. A good example of this is the model of classical physics. It has great success in many respects and yet has no knowledge of quantum fields. But this amounts to saying that it misconstrues the entire set of things it purports to be dealing with! One may suspect that quantum physics also entirely fails to model central aspects of the things that it deals with. But in any case all this model-making plays out in the internal theatre. And as soon as we want to do something else, such as pay attention to our friend, or our child or parent, this model-making approach has to be almost entirely abandoned because it is so useless. A different kind of attention is required.

To approach the world with a religious sense of reverence involves the desire to pay attention *both* in the first kind of way *and* in the second kind of way. The second way is not adopted as a replacement of the scientific enterprise, but as way to open ourselves to the world we are really in, as opposed to any model of it that we may have formed. It is our aim to be fully alive and responsive. God is encountered in the actual relationships which form us and whose sum is beyond us. God is not encountered in the internal theatre, because the internal theatre contains only abstractions and tools.

In that final comment about what I have called an 'internal theatre', I mean the scientific theatre: the one formed by our intellect and where we peruse the assembled items, as we ourselves stand one step removed and act as assessor. We each also have another form of inwardness, and that other form is different: it is where deep things about who we are have their being, and where we do not stand in judgement but can only be. In this other kind of inwardness we are not assessor at all and we are fully present; or at least we are present insofar as we are able to be honest and not distracted. In that kind of presence (I mean the presence of ourselves, not wandering off elsewhere in our distracted thoughts), actual reality (as opposed to any model-like construction) is available to be encountered.

The older part of the Bible describes a journey away from, not towards, the attempt to construct models (images), whether mental or physical, of that *fullness*, unbudgeable fact, lively challenger and holder-of-value which, according to the Biblical witness, defines or declares what all things are. When a Jewish rabbi adopts metaphors such as Father, Shepherd, Judge for this absolute reality, he is indicating avenues of human experience along which truths about God can be conveyed to us. He is not inviting us in to the very idolatry which the Jewish experience so painstakingly abandoned over its long history. I don't have space here for a complete starter-course in Abrahamic theology, but these comments may help the reader understand that intellectually serious theism is not about the sort of Big Brother deity figures one encounters in atheist polemic. And nor is it about the appalling pettiness and unfairness attributed to God in some versions of commitment calling itself Christian.

To be clear: the combination of reasonableness and religious sensibility is not about forming beliefs that are unwarranted by experience. On the contrary, it is about *paying attention*, paying deep attention, to the whole breadth of what informs our experience, including what we inherit from the past and what we learn in the present. It is not about unanswerable questions and invisible worlds; it is about these rocks, these bones, these people, human relationships in the here and now; about evidence-based medicine and good practice in democratic governance; about theatre and art and computer programming and biodiversity and technology and poetry and music and machine learning and architecture and justice and adventurous trust not fear; it is about *being the change we want to see in the world*. It is about the wish to *choose life*, to live in touch with reality in its fullness, not as prisoners of the assumptions and delusions of the age.

Scientific analysis invokes models which are not themselves the natural world, but which help us understand. Religious sensibility involves a willingness to encounter the world directly, interested to apprehend, as fully and truthfully as possible, the nature of each thing and person. We are not always looking for explanation; we are looking for the meaning of the verb *to be*, and for the personal resources to enact our own role appropriately.

To recognize that the cosmos is a creation and a gift, and that our life is lived on multiple levels, and flows ultimately from God and to God, is to engage in straight dealing, to live in the world as it is. But this is subtle, of course. It includes a response at a deep level, a level which most people will not be able to put into words very well. This accounts for the large amount of stumbling words that people employ. A few humble statements about gratitude and forgiveness may in the end be as apt and truthful as any lengthy philosophical discourse, but that kind of simple speech, valuable as it is, leaves us wondering whether or to what extent we are talking in metaphors. Or if we are using ritual then it leaves us wondering what the ritual is saying or what the symbols are symbols of. So we do need some attempt at more sophisticated theological language, and that is what is developed, slowly and carefully, in a worldwide exchange of ideas amongst the deeper practitioners of the various traditions (while the shallower practitioners make a lot of noise in their strivings). A large element of theism is the finding that human beings discover their identity most fully and correctly by recognizing themselves to be in relationship with the deepest reality, whose relation to natural phenomena is not easy to analyse or even express in words, and whose encounter is not adequately described as entirely impersonal.

That was a fairly long sentence. I tried to shorten it but I couldn't, because we are in difficult territory here. Of course we are. As the poet Rainer Maria Rilke said in a letter, 'Things are not all as graspable and sayable as on the whole we are led to believe; most events are unsayable, occur in a space that no word has ever penetrated'.[1] When trying to put into words that which stretches beyond the boundaries of our very language and thought, you can have falsehood or difficulty; there is no third option. But I will try to clarify the *relation to natural phenomena* which I mentioned just now. That relation includes making it possible for physical things to be what they are; affirming the truth of what they are; and allowing to persons the opportunity for growth in spiritual and moral stature through a meeting on the level of their personal identity (as well as on other levels). Let us call this a 'carefully stated' theism. There is also, in much popular or folk religion, a further pair of ideas which can be seen as failed human attempts to contain what cannot be contained. On one hand there is the amorphous spiritual force idea, and on the other the supernatural entity idea.

The amorphous force idea is trying to avoid anthropomorphism but too easily it is treated as a kind of universal personal accessory, one that reassures us that we are spiritually deep but does not make much in the way of ethical demands nor hold out any prospect of forgiveness. This does not engage us on a moral level, and thus it does not have the quality of reality, of that which we must face up to.

The other common folk-religion idea is to conceive of God as a supernatural entity, regarded as an authority figure to be implored, or a fixer to be summoned in to deal with problems or fill gaps in knowledge. This incompletely thought-through version is how theism is often conceived by atheists too; and they are quick to discover its lack of credentials. But this approach has been resisted by all mainstream religious traditions.

---

[1] R. M. Rilke, *Letters to a Young Poet*, (Penguin Classics), letter dated 17 February 1903.

To understand what is going on, it may be helpful to heed the thirteenth-century philosopher and theologian Thomas Aquinas. He serves as a helpful guide to what happens when philosophy in the ancient Greek tradition of Plato and Aristotle meets Christian theology and mysticism. By mentioning Aquinas I don't intend to emphasize one individual, but rather to make contact with the depth offered by a range of commentators, including others such as Meister Eckhart, Teresa of Ávila and the founders of the various monastic movements. We find in them also a resonance with some of the themes of Daoism, Buddhism and Hinduism. Modern people tend to assume that a long time ago people had in mind the simple 'supernatural entity' sort of picture when they spoke of God, and it is only in modern times that we have become more sophisticated. This is wrong. The Hindu and Jewish traditions arrived by the axial age (a period some hundreds of years BC) at a considerably more thoughtful and deep position. And, like Buddhism and Daoism, they are well aware of the impossibility of rendering wisdom altogether into words. The ancient Jewish temple in Jerusalem had, quite deliberately and thoughtfully, a space not a statue at its heart.

What we now call Christianity is a variegated human tapestry, not all of it very well connected to its origins, but the spring it sprang from is a threefold combination of Jewish, Greek and original influences—the third component coming from Jesus of Nazareth in his attitudes and assumptions and style of teaching, a style which is sometimes direct, and often indirect and Zen-like, as occasion demands. Meanwhile, Greek philosophy invites us to try to define things and pin them down; this can be very powerful and it is one of the tools of clear thinking. Thomas Aquinas lived at a time when the Greek analytical approach was being rediscovered and people were working out how it interacts with the mystical end of religion, where language falls silent. The solution is that instead of looking to pin down God as an end point in increasingly sophisticated sets of scientific ideas, we learn to acknowledge that God is prior to any set we can define or generalization we can make, including the one expressed in this sentence! This is why the notion of an 'entity' fails to name God, even if we add the adjective 'supernatural'. An entity, supernatural or not, would be a further item in the set of all items. But God is not that, because God is prior to any set.

This is somewhat like the way a musical chord is not a further melody in the set of all melodies, and friendship is not another legal contract in the set of all legal contracts, and to be merciful is not to abandon justice but to transcend it. So instead we learn to use a variety of names which set up resonances in the human heart, and we see that reality takes us in hand in various ways: sometimes in the manner of a shepherd, sometimes in the manner of a judge, sometimes in beauty, sometimes through the trials of pain, always in solidarity.

To approach God as primarily a provider of answers and presents is comparable to approaching another human being as an intellectual puzzle merely—an object to fill a gap in one's own intellectual or emotional agendas. When we do this with respect to each other, we utterly fail to encounter each other. When person A approaches person B purely with a view to getting their vote or their support, or just in order to measure them, then, for all that a verbal or other exchange may take place, person A never really meets with or encounters person B at all. Person B here illustrates the way in which some aspects of reality are real enough but utterly inaccessible to us if we do not take the right approach.

Making this mistake with regard to God is a fundamental error. It gets you precisely nowhere. Its incorrect and misleading nature has been repeatedly emphasized both by theologians and by the mystical tradition in all its forms, including Hindu, Buddhist, Christian, Sufi Islamic, ancient Hebrew, etc. So if by atheist we mean merely the rejection of the supernatural-accessory version of religion, then all these religious traditions are atheist. But this is to make merely a baby-step in theology. What all those religious traditions want to say to parts of modern-day atheism is, please will you stop being ignorant in your attempts at theology and listen more carefully to what we have to say. (They also would make this same request to the shallow end of their own religious communities.)

I would like to acknowledge that a religious sensibility does not always evoke the name of *God*. Any morally coherent sensibility will correctly be wary of the ways in which God is spoken of— many of them wrong, some of them appalling. But I would like to say a little about how a willingness to recognize God can work out in practice when it stays morally coherent. A carefully stated theism is also a thoughtful theism.

A thoughtful theism, one that has embarked on the intellectual discipline that such thoughtfulness requires, is not easily described and it is not about seeking an accessory nor is it about blame or manipulation or airy philosophy or desperate pleading, nor is it about any objectified description that can be put down in a book. It is much more about living a certain way and discovering an endless well. It turns out to be, to a large extent, an ongoing attempt to learn the nature of reality and act accordingly. The practical consequence consists largely in increasing one's confidence that acting well (i.e. in agreement with conscience and wisdom) can be trusted—it will work out in the long run, whereas the tendency to seek immediate security will lead one astray. This involves developing sensitivity to, and appreciation of, the innate nature and interests of each thing or person encountered. In addition to making scientific investigations into the structure of physical processes, theism also wishes to discover what has been brought about by those processes—what the things that exist *are*—what is their role, what do they offer or signify, and how should we respond to them? One learns to ask, more and more, or with increasing penetration and depth, *what does 'being' and the verb 'to be' signify?* (After all, in Jewish and Christian thought, that verb is almost the very name of God.) In this attempt, all the contributions of scientific analysis can be fully welcomed and used, because they tell us about the being of the physical world. This contingent world is not the same as the non-contingent which supports it, nor does it offer anything but limited clues to the non-contingent, but what clues it does offer are welcome. And one of the clues is the experience of what it is to be a subject and not just an object—one who has felt experience and who encounters other subjects in mutual recognition and regard. We thus encounter a realm of what is so—of being and the verb *to be*—which is concerned with the recognition which one person has for another, and there is hardly anything which this recognition does not touch in one way or another.

A major item touched on by this experience is the question of what kind of person we each aspire to become. The major cultural traditions of the world agree on many aspects of this. When Aristotle wrote of courage as a desirable quality that steers a path between rashness and cowardice, it is hard to find anyone who would disagree. His ideas on straight speech are like those of Jesus of Nazareth and I don't doubt that ancient Norse peoples and modern business people would agree. No one likes a man with a slippery tongue. And so on. But does this mean that we understand all this perfectly well and have little to learn about virtue? Not at all. Our experience here is reminiscent of our aesthetic experience: we perceive something that is beyond us, and we are in the position of reaching for we know not quite what, but it is supremely beautiful and consequently attractive, yet daunting. This asymptotic attractor is what the word *God* refers to, to people steeped in various religious traditions that reach back thousands of years and have roots in many parts of the world. What can be truly said of God is that truth which all human life struggles to articulate: that which our concepts of love, justice, creativity, fairness, courage, scientific insight, wisdom, irony and humour are ever enlarging to embrace. These fill out the *name* which we pray may be hallowed and celebrated.

The turn of the heart, or the will, or the gut—the whole human being, however you want to put it—towards the eternal is a characteristically human, natural, creative and reasonable turn, though also a puzzling one. It embraces the notion that this truth is not entirely impersonal and amoral, nor is this merely about abstract thought: it is much more about experience and behaviour. A widely reported aspect of human experience is that contemplation can open out onto empathy, mercy and the kind of challenge which a good parent offers to an adult child. Such experience is not entirely impersonal, which is why we may say *our Father* rather than remain silent as one does before an artwork. We acknowledge that this is a metaphor, but it is the personal not the impersonal metaphors which come closer to the truth here. Or, one may say that our own attempts at parenthood serve as metaphors for the larger relationship. And also we find that it is fun—liberating—to give thanks in personal terms, and to be able to grumble, and to be forgiven. Faith then rises to the challenge of *behaving* as if the arc of history bends towards justice—a justice which we do not invent but learn. Our prayer becomes *may things be done on Earth as they are done when truth and love prevail.*

I can only use the word 'God' with some hesitation, because I know that I am liable to be misunderstood. I intend by it mainly an attitude of receptiveness to what Jesus of Nazareth was saying and doing, but this receptiveness itself needs to be trained by contemplation of long strands of human history and of the present moment. In particular one draws on a worldwide wisdom tradition and a meditative witness which suggests that those aspects of reality that we name by words such as empathy, respect, justice, courage, mercy, love are not mere offshoots of physical processes, but reflect a unity and creativity which is able to encourage, mould, forgive, require, aspire and affirm.

A willingness to acknowledge that there is, in the reality that holds us all, a sense of *thou*-ness and not merely *it*-ness, is, or can be, a profoundly human and humane willingness. One slowly becomes aware of a calmness deep down things, a sense of being loved but not taken too seriously. This can gradually transcend whatever has been unjust in our background, giving us the resources to avoid bitterness and instead find endlessly renewed ways to meet one another more fully, equally, and without fear.

One of the important disciplines here is *epistemic humility*, which is the practice of being cautious and realistic about what degree of knowledge or certainty one can claim to have. Nothing is certain here, except that we have valuable lives to live and what will survive of us is love. The intellectual apparatus developed in careful long-term theological discussion points a way in which the experience of a religious response in the heart can be received with intellectual integrity and welcomed in good conscience. Theism as a way of life involves honest appreciation of one's gifts and finding ways to put them to use in the service of good; theism as a metaphysical claim is first the claim that the verb 'to be' signifies, in the end, a degree or level of being which the components of the natural world do not in and of themselves attain, but express aspects of, and second that this fullness of being can be encountered by persons through every aspect of their existence, including their very personhood with all its hopes and regrets.

Sunbiggin Tarn

The ground will only be holy if you feel like taking off your boots.

It starts at home when you weigh up what can earn its place
in a pack that has to be carted eighty miles. Tent,
coat, knife, sleeping bag, spare shirt, water bottle,
map, short-handled toothbrush.

It's no good tripping up in a car and rolling out picnic blankets.

Slog over Begin Hill and familiarize yourself with
the peat of Ravenstonedale Moor; the rocks of Hard Rigg.
Come at it late in long afternoon sunlight,
weary, looking for a place to pitch.
Then in silence under no man's gaze
Be there.

The breeze stirring the heather, the tarn
somewhat below and a little ways off,
the light darting across the landscape
and catching you unawares,
suddenly full and seeing,
earthed and aching with life,
hungry, enlarged
and made smaller by a
still opening.

Let it be not a find, but a gift.
Let it be not a taken, but a token.

If you drive there you will find only a lake,
and birds to watch; a picnic spot,
a pleasant afternoon. There is no
vehicle that can carry you to where
you can only be.

CHAPTER 17

GREAT IS THE POWER OF STEADY MISREPRESENTATION

The next few chapters of this book are going to be somewhat more edgy. I have already made some critical remarks about intellectual practice in the public sphere, but I have tried to keep the tone light. I am now going to suggest that some of the practices I have critiqued are not just forgivable overenthusiasm or honest mistakes, but propaganda, and contributors to an attempt to use the levers of academic power to perform an act of cultural imperialism. By cultural imperialism I mean the attempt to put forward subjective and disputed views through techniques such as the assumption of an authority one does not have, and partial or unbalanced presentation of the intellectual territory. When this is done forcefully, so that another is not free to see a complete or balanced picture and exercise their own judgement, then a form of propaganda or imperialism is on show, and ought to be called out as such.

The title of this chapter is a well-known quotation from Charles Darwin in the sixth edition of *On the Origin of Species.* It was an apt comment on those who had misrepresented earlier editions of

Liberating Science. Andrew Steane, Oxford University Press. © Andrew Steane (2023).
DOI: 10.1093/oso/9780198878551.003.0017

that work.[1] The comment has been quoted more recently in order to show in its true colours work that passes itself off as the promotion of critical thinking, but is really an attempt to undermine the teaching of well-thought-through scientific ideas (especially Darwinian evolution). The comment can also correctly be applied to modern-day attempts to connect standard evolutionary biology with atheism in an exclusive relationship, as if the former implies the latter. The fact that the word 'evolution' has become synonymous in a lot of people's minds with the word 'atheism' must stand as one of the most successful public misinformation campaigns of modern times. If 'success' is the right word to use.

Let me try to get the aim of the next few chapters clearly in view. The critical remarks I shall make about contemporary atheism are not addressed to the whole of it, but to one part. However, it is the part most visible to the culture at large, through magazines, lectures, videos and popular books. And what I am seeking is reform of the mainstream scientific community. My aim is that the latter should not endorse the attempt by this part of atheism to claim for itself unique access to scientific credentials. People will continue to make claims, but the community as a whole decides whether to acquiesce in those claims, and if the claims are illogical or misleading or unjust then they ought not to be accepted.

Those with scientific training have a welcome duty to hand on the practice and understanding of science to the next generation, and they mostly do this thoughtfully, generously and well. But education of all sorts involves a certain attitude of graciousness on the part of the teachers, and a willingness to be well-informed

---

[1] By this remark (on page 421) Darwin was underlining his previously stated opinion that natural selection was not the exclusive means of modification of biological species.

about human culture in general and about one's direct audience in particular. Therefore it involves that science teachers, whether at school or university level, should resist the suggestion that religious commitment across humanity is neatly divided into atheism and the rest; and they should resist the idea that high-quality, genuine science belongs to atheism or is somehow not as truly owned or shared or celebrated or developed by people who avow the kind of theism I described in Chapter 16. What I think is happening at the moment is that editors, custodians of scientific societies, and the undercurrent of informal banter which shapes the turn of phrase you encounter in public lectures, have drifted into a form of prejudice. The prejudice is to perceive a diverse thing as a uniform thing, and to describe a whole using adjectives or observations that are only true of parts.

This prejudice is not, at present, anywhere near as damaging as many other forms of prejudice, but I think that if it persists then it risks becoming very damaging, because it will exclude large groups of people from scientific careers and risks leading the very practice of science astray.

I would like to clarify what I mean by perceiving a diverse thing as a uniform thing. Suppose, for example, that someone takes the view that left-handed people are gullible and not quite as committed to rationality as right-handed people. This hypothetical prejudiced person will no doubt be able to find examples of left-handed people who are gullible, and of right-handed people who are reasonable. It is a not a prejudiced statement to say that there are left-handed people who are gullible; it is a statement of fact. The prejudice comes in when it seems that this is the only fact we ever hear about, or when someone never sees or admits to themselves the full picture about left-handed people, or claims that left-handedness is itself liable to produce a substandard quality of intellect or of intellectual contribution.

The prejudice at large in contemporary culture is to invoke a group of people called 'the religious people' and then among this group to highlight examples of faulty reasoning or bad grace and then to generalize from this to the group as a whole. As evidence of this prejudice in the mainstream scientific community I could allude to many examples from my own experience, and from scientific magazines, lectures, and widely read popular books. But I admit I do not have a properly researched body of evidence to display, and I am reluctant to list the examples I have experienced. Maybe this chapter should be seen as a cry of pain. What I hope it is is a way to open up the scientific community and allow it to reconnect with the generosity of spirit which is its true heart.

That heart will, I think, be able to appreciate that the modern-day phenomenon of young-Earth creationist theology in a significant fraction of evangelical Christianity in the USA is in part a gross mishandling of the Hebrew and Christian inheritance (the Bible in particular) and in part a reaction to the way biology has been presented. In the 1880s ideas claiming support or motivation from evolutionary biology sought to divide up the human population along lines of class, race or ability and prevent some groups from having children. You can find pictures of placards saying that the poorer classes ought not to be allowed to 'breed', and the people holding the placards are just ordinary people who think they are being scientific. The label of 'science' was also attached to some of the worst atrocities of the twentieth century. This was not properly practised science but an abuse of it. But when abuses are as severe as this then it is not just understandable but right that people should be wary of what world-views may be being packaged up under the label of 'science'. Throughout this period, and right up to the present day, there remains an urban myth calling itself 'the Darwinian view of life' which conjures up 'a panorama of brutal struggle and constant change'.[2] I call this an urban myth because in fact the process uncovered by Charles Darwin, Alfred Russel Wallace, Gregor Mendel and large amounts of further research is not correctly summarized as a panorama of brutal struggle, and its relationship with change is rather interesting. In fact, it is a process of remarkable stability along with a positive capacity to adapt to change. And it is not so much brutal as simply what it is. Insofar as it is a mechanism, it will be mechanistic. Insofar as it includes organisms which have the capacity to recognize others, it also includes the capacities which go with that: decision-making, negotiation, cooperation, conflict and conflict resolution. Predation amongst insects is largely mechanical; predation amongst mammals is not. Predators and prey extend to each other a kind of wary respect as they live alongside one another. And as they try to outwit or outrun one another their awareness of both their own and the other's situation is whatever awareness their physical nature can furnish—not the same as human awareness, but not so different as to be entirely impossible for us to guess at. A reasonable guess is that non-human mammals, for example, live quite richly, marshalling a plethora of smells, tastes, sights, sounds and touches. Insects, meanwhile, make large and necessary contributions to the ecosystem and also cause a certain amount of pain and frustration to mammals and birds, but they do not delight in that

---

[2] Cary Funk, Greg Smith, David Masci, 'How Many Creationists Are There in America?', *Scientific American*, 12 February, 2019; https://blogs.scientificamerican.com/observations/how-many-creationists-are-there-in-america/.

pain. Nor does the fox delight in the pain it causes the rabbit. Overall, life on Earth continues to be what hunters and gatherers and farmers have witnessed it to be ever since the first cave paintings: full of patience, courage, wonder and beauty alongside pain and distress. It is not the 'arena' or death-struggle which some people like to suggest. I will return to this in Chapter 19. Here I am commenting on the way evolutionary biology is presented to people.

An interesting study by Pew Research Center was reported in 2019.[3] Two ways of framing questions about evolution were tried out on members of the American population. Let's call them frames A and B. In frame A, surveyed people were offered the single question, 'Which statement comes closest to your views?', in answer to which they could pick among the three responses: 'Humans have always existed in their present form'; 'Humans evolved; God had a role'; or 'Humans evolved; God had no role'. In frame B, surveyed people were asked 'Which statement comes closest to your views?', in answer to which they could pick between the two responses, 'Humans have always existed in their present form'; and 'Humans evolved'. A second question was then addressed to those who picked 'Humans evolved', and it was 'Which statement comes closest to your views?', with the options 'God had a role' and 'God had no role'.

Overall, then, frame A and frame B end up offering the same three options.

For most of the religious groups who were polled, respondents opted at levels around 70% to 90% for the notion of human evolution, whether the questions were framed as A or B. Among

[3] Ibid.

the two groups who were not so comfortable with evolution, the responses were very different in the two frames. When polled according to frame B, the group identified as 'white evangelical protestant' answered 66% for 'have always existed in their present form' over the alternative offered in frame B ('humans evolved'). But when the same group was offered frame A they mostly picked the evolutionary response out of the three offered in that choice: 62% for 'Humans evolved; God had a role'. A similar difference was reported for a group identified as 'black protestant'. Obviously all these labels, and the questions themselves, are very crude. The interesting part is the way the responses changed. What it shows is that when the question was purportedly about biology, it was not so interpreted by the respondents. They responded as human beings whose awareness includes a political struggle and who want to signal their commitment to God if they can. The question 'do you think humans evolved?' was, to many of them, a proxy for 'where do your loyalties lie concerning education and religious commitment?'. As soon as they were offered a way to signal that commitment while picking 'humans evolved', many of them did so. This is not to suggest that they thereby opted for a nuanced theological position that is altogether consistent with science; quite likely for many of them that is not their position either. But the question I am looking at in the present chapter is one for the mainstream scientific community and its guardians. It is: *what is the best way forward to promote science education in this situation?* I think that just alluding to tired and faulty arguments offered by popular preachers (and a few scholars who should know better), without even mentioning the high-calibre discussions that exist in the philosophical and theological literature, is counter-productive in educational terms, and promotes prejudice.

CHAPTER 18

FRUIT PIE

O nce upon a time there was a country of farms and hills and valleys, and in one of the valleys there were three villages.

In the first village there were many orchards of lemon trees. In the spring-time the blossom of the lemon was beautiful, and the villagers were very proud of their trees, though the fruit was difficult to eat.

In the second village the people kept honey bees, and they were partial to all things sweet.

In the third village they grew wheat and made bread.

The people of the three villages had enough to eat, but they were quite poor and their diet was rather bland. They had some livestock and a few crops and fruits, but they had not developed much in the way of ideas for cooking food.

One day, after studying food and cooking for a long time, a man called Mr Dearwin discovered a way to make fruit pie. He wrote it all down and published it and it caused quite a stir amongst the three villages.

In the first village the new recipe was welcomed with great enthusiasm. They made the fruit pie according to the recipe which Mr Dearwin had discovered, and then they enthusiastically added lots of lemon juice to the mix, and they put bits of lemon

Liberating Science. Andrew Steane, Oxford University Press. © Andrew Steane (2023).
DOI: 10.1093/oso/9780198878551.003.0018

peel on top, because they were proud of their lemon trees. Then the lemon villagers brought their pie to the market to sell.

The inhabitants of the second village were a bit more conservative in their tastes, and they were not sure about the new idea, but they were willing to give it a try. So they bought some of the fruit pie from the lemon village, and brought it home. But when they got it home they found it terrible. The juice was sour and the bits of lemon peel on the crust were dry and bitter. It was just disgusting. You couldn't possibly eat it!

When they heard this, the lemon villagers were at first disconcerted, and then indignant. 'How stupid those honey people are!' they exclaimed. 'Can't they see what a great recipe Mr Dearwin has discovered? It is pure genius. It has revolutionized our understanding of cooking! From now on all cooking will be informed by Dearwinian ideas.'

When they heard that they were being jeered at in this way, the people from the honey village felt that the fruit pie must be partly to blame, and anyway it tasted horrible. So they mostly decided to keep away from fruit pie.

But some of them decided to try cooking fruit pie themselves. A group of the bee-keeping villagers followed the recipe from Mr Dearwin, and then they added to it. They added extra sugar to the mix, and they also poured honey all over the top. Then they liked the pie, and they brought it to the market, congratulating themselves on how clever they had been.

The lemon villagers took one taste of the extra-sweet fruit pie and they hated it. 'Yuck, it's so sweet!' they said. 'You can't taste the fruit at all! Those stupid honey people have totally failed to understand the whole point of fruit pie!'

After this there was a great squabble about how to make fruit pie, and especially about how to teach cooking to children. The two villages set up committees and laws and rules for teachers, and had many arguments.

Meanwhile the inhabitants of the third village read the recipe just as it was, and added neither honey nor lemon, and they did not need endless squabbles about the teaching of cooking. Eventually they sent along a representative to give evidence at the never-ending court case run by the other villages. 'The bee-keepers are making the pie wrong', said the representative, 'but it is not because they can't or won't understand cooking; it is a reaction to the peel which the lemon-keepers introduced. If you get rid of that then the ingredients will be fine.'

CHAPTER 19

CONTEMPORARY THOUGHT AND EVOLUTION

Of the range of theistic opinion in the world, some is in logical opposition to major parts of standard scientific knowledge, and some is not. The latter (i.e. the part of theism that is both logically consistent with, and welcoming towards, standard scientific knowledge) is a mainstream part. For example, as I remarked in previous chapters, most theologians accept that whatever view we take of the physical universe has to be consistent with evolutionary science.[1] What I want to argue now is that science education is significantly hindered by attempts to deny this, to ignore it or to hide it from young people.

The natural world is morally and aesthetically ambiguous. It can amaze us with stunning beauty; inspire and humble us before the courage shown by many creatures; and also horrify and repel us with wasting disease and the mechanical processes of fire and flood, the inexorable logic of parasitism, the brute facts of trauma and child mortality. Standing in the midst of, and a part of, all this, one has no easy answers to questions about ultimate meaning, but one aspect at least is welcome. One may find it both welcome and fulfilling to know that we are cousins with

---

[1] K. Ward, *A Vision to Pursue: Beyond the Crisis in Christianity* (SCM Press, London, 1991), and c.f. the Clergy Letter Project: http://www.theclergyletterproject.org/.

Liberating Science. Andrew Steane, Oxford University Press. © Andrew Steane (2023).
DOI: 10.1093/oso/9780198878551.003.0019

all other living things. For all the pain and struggle of life, one may find it profoundly beautiful that living things are products of improvisation and gradual exploration and epic perseverance. And one may appropriately feel honoured, not ashamed, to be a part of this story. This is, at heart, why evolutionary science is unthreatening to a carefully stated theism.[2]

But if it is presented as threatening, then people will feel threatened.

Let's suppose that our aim is that the majority of the population should be reasonably well educated about biology, so that they can enjoy its understanding and also develop sound judgement as citizens and voters. The way to achieve this is to show the processes and the evidence and the way it all hangs together, and to invite a communal response as to what it suggests about ourselves and our place in the world—but not to try to dictate that response. This is broadly how science contributes, and how its greatest practitioners, including for example Charles Darwin, approached the task of education. But in recent decades there has

---

[2] One might also remark that situations in biology can be read in more than one way. Upon witnessing a seagull on a cliff gobble up the eggs of its neighbour when the neighbour is away, one person will infer that this is what all the other seagulls want to do 'really', while another person will reflect that most seagulls do not do that and therefore probably do not want to do that. When considering the attentive behaviour which goes on between monkeys in a family group, one person will describe it as a robotic response like the meshing of cogs in a complex machine; another will judge that that metaphor is inadequate. On witnessing the great height of trees in a forest, one person will call it a waste of effort, while another will call it an opportunity for an ecosystem. And so on. Contemporary research in biology shows us more and more that mutualism quietly permeates the natural world. The human body is itself a symbiosis. Parasitism is, by far, the exception not the rule. And when we discover the strategies by which plants and animals grasp opportunities to live, and in this compete, we discover an intricate network which excites our wonder and marvel alongside our empathy and dismay when individual creatures suffer. Its study is fulfilling like the Olympics, not demeaning like bear-baiting. And we can be reasonable about the metaphors we bring to it. The life of plants and insects is morally neutral. That of mammals and birds has the stirrings of ethics. The morally objectionable arises precisely in proportion to the morally responsible and to freedom itself. This has been discussed at length by many writers; I have offered some reflections in chapter 17 of *Science and Humanity*.

been another approach, in which writing about evolution is spiced up by bringing on a straw man here and there, or by enjoying the frisson of proposing that Darwin's idea is not just a good idea but a 'dangerous' idea. But the straw man always has a religious hat on, and Darwinian evolution is said to be 'dangerous' to theism but educational to atheism (after all, how could it be educational to both?). These are the tactics of the propagandist, and the result is anti-educational to all parties. To the (large) in-group which enjoys this approach, it reinforces stereotypes about the irrational 'religious people' and prevents that group from seeing their neighbours for who they really are. To those religious groups which are nervous about evolution, it fails to offer a constructive way forward. But such a failure is to fail utterly in educational terms.

The message, rarely stated openly, but plainly implied, is that the only alternative to brave and nimble atheism is some intellectually second-rate version of theism.[3] Fun as it may be to its practitioners, this kind of approach is not educational and just succeeds in raising barriers to understanding for young people who have not deserved such misleading messaging. I will not be able to persuade these practitioners of stage-magic to desist, but that is

[3] The conservative creationist versions of modern-day Islam and Christianity do have to be deemed intellectually second-rate when they fail to perceive the rich use of symbol and metaphor in ancient literature and when they fail utterly to grapple with the large body of scientific knowledge about biological processes and evolutionary history. Intermediate between this and the mainstream scientific work is the body of ideas which have been promoted in recent decades under the names 'irreducible complexity' and 'intelligent design' (mentioned already in Chapter 13). There is a complex combination of social issues here. The 'intelligent design' movement has among its stated aims that it proceeds in a scientific manner, following evidence and inviting reasoned debate. Its representatives have acted with commendable civility in the face of insult. However, reactions have been strong because the movement is suspected of being disingenuous and in fact unscientific under a scientific gloss. The writing in support of ID that I have read has shown, in my opinion, too great a readiness to bring in unwarranted conclusions. On this basis it has been critiqued by various reviewers, of whatever background as regards religious commitments.

not my aim here. My aim is to ask representatives of the scientific community, such as editors, boards of education and custodians of scientific societies, to aspire to a higher level of scholarly standard in its public voice.

The attempt to link evolutionary science exclusively to atheism fails in logic. For this reason alone, it ought to be resisted by the scientific community. Furthermore, the important and fascinating evolutionary story of life on Earth is part of the intellectual property of the entire human race. It does not belong exclusively to any subgroup, and the contemporary attempt to claim sole ownership of it by one subgroup is objectionable on the grounds of ordinary justice. That attempt typically takes the form of a presentation of some facts about what happened, with a coherent argument about the natural process involved, followed immediately by drawing a contrast with ideas drawn from certain naïve or self-absorbed versions of theism, as if those were the only or the main versions to be found in the world. This kind of journalism and lecture-hall showmanship presents at best ignorance of scholarly discussion of the interplay of science and religion, and at worst propaganda. It represents, therefore, an abuse of science, and this should be made clear to the general public.

Popular writing, film and television drama rarely progresses beyond the most banal conception of what long-standing religious vocabulary could possibly be dealing with. Writers make an effort to invoke sophisticated treatments of technology and the physical world, and then lazily summon up a troop of stereotypes and ignoramuses brought on to inhabit the 'faith zone', fit only to be smirked at or swept aside by the brave protagonists.

In fact, the majority of academic theistic thinking, both historically and in the modern world, is at pains to point out the subtlety of the possibilities when it comes to stating the relation between the natural world and its ultimate grounding and purposeful trajectory. The situation is as subtle as the mind–body problem, for example. It is not the same problem, but it is equally subtle. It has nuance and resists analysis and complete consensus. This is related to the fact that it is chiefly about interpreting what phenomena mean or express or amount to, rather than describing physical causes and structures.

Now it may be objected that if you count members of the general population having some version of theistic belief, then you will not find, on average, a very sophisticated version of that belief, insofar as it concerns abstract metaphysical questions. But the same can be said of the average atheistic belief. In the exchanges that take place across the tables of pubs, wine bars, and internet chat rooms, the average level of discussion is not high, no matter what kind of religious or irreligious position people are speaking from. So the truth of these matters is not found merely by cutting down the straw men offered by various popular but simplistic arguments, whether they be in atheist or theist versions. And they do come in both versions. The level of theological insight on show in blog and YouTube polemic is painfully low. The treasured combination of creative understanding, value and possibility that underpins the universe and gives hope is hardly even glimpsed or mentioned; instead the name of 'God' is attached to a fantasy figure, one written of with half-baked attempts at seriousness but really as banal as a horoscope. This is dangerous because the end result is prejudice: the prejudice that prevents one person from seeing or hearing another as they really are, because the first person has both refused to learn another's language, and also refused to admit that this is so. We end up with the self-contained man or woman, comfortable in their own imagined maturity, thinking that rejection of the imaginary deity-figure is some sort of end-point in theological maturity, when it is really just the tottering of a toddler—valuable in its place, but hardly the final aspiration of a fully functioning adult with a wise mind and an open heart.

Even if many people do entertain a rather anthropomorphic picture of God that a more reflective person could not endorse, we should not be too quick to claim superiority, because our own view of our ultimate context almost certainly fails in equally profound ways. I think what matters here is the quality of someone's reasonableness and the quality of their behaviour, and we may form legitimate opinions about those. But of the two mistakes of saying that a cactus plant (for example) is entirely unlike a machine and of saying that a cactus plant is a machine pure and simple, it is not easy to say which is the more mistaken. Or suppose that two people have little knowledge of elephants, such that person A thinks that elephants are small, furry living things which hop on two feet and have a small nose, a long tail and no tusks, while person B thinks that elephants are large, grey robots with tusks, a long nose, no fur and a short tail, standing on four feet. Person B here has got a lot of facts straight but arguably they have missed the crucial fact. Person A, for all their mistakes, at least has understood correctly that an elephant is a living thing.[4] It is similar with statements about the nature of reality, or what

---

[4] The distinction between a living and a non-living thing is a hard philosophical question; I speak of it here in a shorthand. By a 'robot' I mean an input/output processor with a very restricted inner experience and a sharp delineation between itself and everything else. Living things, by contrast, connect seamlessly to a flow and a network of relationships of which they are a part. For a recent lengthy exposition of this, see McGilchrist's *The Matter with Things* and (more briefly) my *Science and Humanity*.

allows the universe to be what it is. If in fact theism is at all on the right track (which I do not claim to know for sure) then even its most unsophisticated versions may have more merit, for many purposes, than the most erudite statement of a purely naturalistic philosophy.

But I digress. The main point for the present chapter is simply that theism need not disagree with atheism about the internal structure of impersonal physical processes. This suffices to prove my claim that the attempt to link evolutionary science exclusively to atheism fails in logic.

Scientifically informed theism and atheism still differ, but not about whether or not bears defecate in the woods, nor about whether or not genetic variation happens in the way the evidence suggests, nor about whether neo-Darwinian ideas correctly describe the process of speciation etc. What theism brings, along with much else, is an enlarged sense of gratitude at the goodness of the world, and grief at and objection to its badness (including one's own badness). These experienced senses, not philosophical argumentation, are the engine room of the human response to God. The lessons of history and the analytical powers of philosophy and theology serve to hone them and protect us from veering into inappropriate guesses about the nature of God. The combination yields a satisfying whole, and this is what makes us willing to use the word *God*, even with all the admitted confusions and misapprehensions that the word has accumulated over the years.

All this shows how it is that religious and irreligious commitments can have scientific study in common while retaining their right to disagree on overall interpretation.

Justice and word-games

It might be objected, in the present context, that if my aim is to tackle the phenomenon of ignorance of biology, and the rejection of well-supported evidence and understanding, and attitudes of suspicion towards well-motivated research, then I ought to address my remarks primarily to certain religious communities such as conservative Christianity and Islam. I agree that those communities are going to have to put their house in order if they are ever to earn respect and not bring their own message into disrepute. However, the responsibility of one community to behave well does not abrogate the responsibility of another to do the same.

The attempt to present science as atheism, or as if it could not also be consistent with theism, has a pernicious effect on science education in America, various Middle Eastern countries, Africa and other parts of the world. It is a major driver of the resistance to science expressed in large parts of the population. If in fact science were the exclusive intellectual property of atheism then one would have to live with and manage this resistance. But since in fact such exclusive claims are illogical and unjust, asserting them is not the right way to manage science education. For all these reasons, the scientific community has a duty to promote intellectually coherent discussion of the relationship between science and religion. That relationship is rich and subtle and by no means a simple dichotomy. Attempts to claim that it is a dichotomy are either incoherent in themselves, or are based on a redefinition of the word 'religion' so as to bring about the correctness of the dichotomy. For example, if one begins by using the term 'religion' to mean, by definition, reliance on authority figures and blind belief, or if one insists that it can only refer to a form of pseudo-science or quack medicine, then obviously 'religion' will be in contradiction with reason and good sense. But this kind of polemical strategy advances neither knowledge nor understanding, but merely degrades the clarity of language and consequently the ability for learning to take place.

This is not to suggest that religion is an unqualified good, which plainly it is not. Rather, it is a human phenomenon which can be practised well or badly and used for good or ill. But the claim often made in the name of atheism, that atheism is the sole or the main champion of reason and rationality, and can fairly look on others as if their reason, or their commitment to reason, was in question, is terrible, horrible. To suspect another's very rationality before they have even opened their mouth to speak is to place them in a subhuman category.

When a person adopts a label such as 'rationalist' for themselves or their group, what work is the label doing other than to assert that anyone not in the group is somewhat less rational, or less committed to rationality? This is simply prejudice announcing itself. It is like those political parties which present themselves as the voice of patriotism or something like that, as if they understand their country's national pride better than everyone else, whereas in practice they often have a skewed vision of it.

My present purpose is not to convince anyone to pick one position or another, but to object to an unjust situation and to show that it is indeed unjust. I mean the injustice of removing the principle of 'innocent until proven guilty' in intellectual matters, and of redefining words such as 'faith', 'God' and 'religion' in an Orwellian manoeuvre, so that people are no longer able to communicate.[5] It is unjust to remove the principle of innocent until proven guilty, concerning the quality of religious thought, because the record will show that, alongside the intellectual dross, there exists a body of high-calibre scholarly endeavour here. Such endeavour is addressed, for example, to showing how language itself works and how networks of ideas adjust to new knowledge. It takes an interest in what kind of thing can reasonably be said to be going on in religious response. This involves careful thought around the nature of truth and the boundaries of language itself. In order to earn the right to comment one has to show some minimal competence with Heidegger and Wittgenstein and Bonhoeffer and MacIntyre, or their equivalents, or at the very least to acknowledge that such thinkers have grappled with the issues a good deal more cogently than any work which adopts towards them a breezy dismissiveness.[6] The experience of personal encounter is managed differently from analysis and reduction; it involves building trust and a voluntary mutual giving of ourselves into the hands of each other, as well as a host of other subtleties that art explores and history recounts. Philosophers such as L. Wittgenstein and S. Weil, theologians such as H. McCabe, D. MacKinnon and R. Williams, are among many who have written insightfully about the nature of religious categories; programmes such as the Gifford Lectures have presented informed discussion of the roles of scientific analysis and correctly motivated religious response.

[5] For example, the word 'faith', which means, approximately, trust on the basis of suggestive evidence and a sense of value, gets treated as a synonym for 'delusion' or 'uncritical acceptance'. This kind of redefinition of a perfectly good word amounts to a deliberate manipulation of language so as to prevent thought and suppress dissent.

[6] I picked examples from western philosophy, but much of the practical outcome can also be gleaned from a thoughtful interaction with wisdom literature from many parts of the world.

Education and its enemies

How is it possible that now, in the presence of so much evidence and good sense in the mainstream scientific account of the development of life on Earth, there remains in the general public such a large amount of refusal to accept it? Is it simply owing to conservative religious habits of mind? I have argued in this chapter that although this may be part of the cause, it is far from the whole of it, and a significant factor has come from scientific commentators who have deployed the conjuring trick of the magician's box (Chapter 13) and have linked our current understanding of biological science to conclusions which are vastly beyond its capacity to declare. I mean, for example, the conclusion that we ought not to embrace life with a sense of personal gratitude, reverence, forgiveness, and universal hope, or else if we do so then we may do this only while refusing all use of personal categories when we respond to experiences of forgiveness, gratitude, dismay, wonder and hope, and as we assess evidence that the universe is on a meaningful trajectory.

Once expressed in careful terms like that, it becomes obvious that the study of biology is not competent, on its own, to pronounce a judgement one way or the other on such conclusions. The fruit pie of variation and natural selection really does not need the lemon peel of a purely naturalistic philosophy. It is not just a little untrue, but far from true, that the history of life on Earth is consistent with, and congenial to, only one type of world view (the one which prefers to be seen as non-religious). The point at issue is not, 'did evolution take place through genetic inheritance, variation, long timescales and natural selection?' The issue is, 'in what forms can human life be lived in good conscience, in view of facts such as these, alongside all the other aspects of our experience?' There are forms of both atheism and theism which offer thoughtful, creative and well-argued answers to this question. We owe it to the next generation to offer fair appraisal of the breadth of reasonable opinion, rather than trying to force them to choose among fortresses of narrow opinion, each perpetuating ignorance of the other.

I will finish the chapter by saying a little about how education in this area can go better. It is not that one pretends there are no differences of opinion; rather one seeks to understand what it is that different repositories of human wisdom have to offer.

The disagreement between a fair-minded atheism and a carefully stated theism is, in my experience, mainly about two things. It is partly owing to a difficulty people may have in grasping that God is not anything they can define in their internal scientific theatre and see as a component in a sequence of cause and effect, and yet certain aspects of what happens in the world are rightly seen as more owing to God than others. This should not surprise us too much, since the same can be said of goodness, and of truth, and of beauty. The representation of reality which we are capable of forming in our internal theatre is incapable of capturing the whole truth of reality itself, of what actually happens in the world. Beauty cannot be captured that way; another person cannot be explained that way. The very act of embarking on such an explanation ceases to allow them to be a person. God, like love, does not explain all things but bears all things.[7]

---

[7] I have tried to be even-handed in bringing in religious elements in this book; I hope the reader will not mind my mentioning that it is the deeper elements of spirituality, like the one expressed in the concluding thought here, that are the deeper refrains of the Bible. This will be missed by anyone whose interaction with that collection of documents has been cursory. The realization is woven through the collection, starting from one of the

The second disagreement between atheism and theism is, I think, to do with showing what is the value of the practices and sayings which are adopted in human efforts to recognize God. Often they are not so much objected-to as found to be unappealing and it is decided that time could be better spent another way. We can agree that this is a fair reaction to much of what goes on. Nonetheless, a lot more silent contemplation, humility and expressions of grateful joy might have done us all some good, in the last fifty years, by slowing down the consumer culture and allowing us to see each other more clearly. They may also have helped us to build relationships more wisely and thus reduce the tsunami of family break-ups which children in the western world have had to live with.[8] Such practices might also have helped us to pay more attention to the natural environment and the long-term impact of our actions.

Whatever one's conclusion about the intellectual virtue or otherwise of this type of perception and behaviour, the disagreement between atheism and theism here is a disagreement not about science *per se* but about how its contribution is best to be nurtured and embraced. Like many other issues in human life (such as aesthetics, jurisprudence, mercy and fairness), this disagreement is not resolvable by the scientific method and therefore it is not essentially scientific in character. It follows that whereas one may make claims about the unique reasonableness of atheism, one may not correctly or logically make them in the name of science.

---

most ancient texts—the book of Job—then periodically re-gathered through the long saga of ancient Israel, and powerfully reasserted in the death of Christ and the new community.

[8] This is not primarily about divorce, but concerns absent fathers and, more generally, children whose reasonable expectation of continuity of care by parents is not met, whether or not the parents were legally married. It is a huge elephant in the room of contemporary western society.

To call such a claim 'scientific' or to present it as a consequence or result of scientific study is an abuse of science.

It is a valid and worthy aim to argue or try to show that one's own metaphysical commitments are consistent with the value and goodness of what science can do. One may also critique another's commitments by arguing that they do not do justice to science. But such arguments and critiques are not themselves part of what science can adjudicate. Hence it is not correct to claim that science directly supports one's own religious or irreligious commitments, as if they could be obtained by scientific argument, or as if one's own position had unique access to scientific credibility. When leading scientific commentators writing for the general public fail to grasp or abide by these considerations, they commit an injustice against young minds because they mislead. It also results in the injustice of ordinary educators being placed in an impossible situation, as has happened in the USA.

Jonathan Sacks[9] described the situation correctly and helpfully when he said that 'science masquerading as religion is just as unseemly as religion masquerading as science.'

I write in hopes of promoting a considered response which may further the cause of science education both in Europe and America and throughout the world.

---

[9] 1948–2020; Chief Rabbi of the United Hebrew Congregations of the Commonwealth 1991–2013.

CHAPTER 20

BRIGHTLAND

O nce there was a village of houses, farms, a school, a church, a blacksmith's, a small college and some pubs. The people owned their houses and gardens, and the owners of the farms owned the fields. The blacksmith owned the forge, and the founders of the college owned the college. So you see, most of the land was owned by someone or other. But in the middle of the village there was a large green space called the Common, and this wasn't owned by anybody. It was the common land. Everybody could go there, and walk their dog and graze their horse or just sit or lie in the sun.

After a while, as the village prospered, people had more time to enjoy the Common. As they walked there, they began to notice more and more of the flowers growing around about, and the trees. Then they started to dig small holes, and they found there were interesting fossils and minerals in the ground, and the roots of the trees. They also enjoyed the flow of the stream and the patterns in the light. They liked to see the ice crystals that formed in the winter frost, and the rainbows in the dew on sunny spring mornings.

The people came to see that the Common wasn't just a place to walk. It was also worth studying. They found out, to their delight, that the Common was marvellous and fascinating. Studying it

Liberating Science. Andrew Steane, Oxford University Press. © Andrew Steane (2023).
DOI: 10.1093/oso/9780198878551.003.0020

was not just interesting but also quite fulfilling. It was also a great way to help people act as equals towards one another, because everyone was free to go there. This is because it was common land. Nobody owned it, as I already said.

Some of the people who loved the Common would spend a lot of time there. They got to be good at studying it. They weren't the only ones good at studying the Common, but they were good at it. So they talked to one another about what they were finding out, and they formed themselves into a club. They called themselves the 'Brights'. They saw the bright future that would come as more and more people joined their club and took up their studies. They had two club rules.

Rule 1. Members shall share their discoveries about the plants and the streams and the stones lying on top of the ground.

Rule 2. There shall be no digging under the ground or climbing in the trees.

The reason for the second rule was that digging under the ground was difficult and often led to places where it was hard to see. No one could agree what was being discovered, so the Brights decided to have a rule against it. Also, they thought that climbing in the trees was not sufficiently serious.

The happy Brights built themselves a clubhouse, and they put up a little stall on the Common, inviting people to join their club. As people went for a walk on the Common, they might meet one of the Brights, or find the stall. They could join up if they liked. There were good lectures to go to, and all the fun of discovering about the plants and the water and other such features of the common land.

As the club grew, the Brights extended their clubhouse. Part of it used up some of the common land. According to the ancient rules of the village, no one was allowed to put their private buildings on the Common, but somehow nobody seemed to notice. Well in fact, a few people did complain, but the Brights just said, 'Those moaners! Don't they appreciate all the discoveries we are making?'

One day, a man came to the Common, wanting to walk there and enjoy studying the plants and rocks, but a member of the Brights took him to the sign-up stall and said, 'Look here, fellow, you can't walk here until you have first signed up to our rules. Here they are.' The man looked at the rules and said he liked the first one but not the second one. He was interested in minerals and roots, so he didn't want to obey the no-digging rule. 'Don't you like the wonderful plants and the water and the stones?' said the Bright. 'Yes I do', said the man. 'I like them very much, and I have studied them all my life. Now please will you let me go and carry on finding things out!' 'No, not until you have joined our club', said the Bright. The man did not want to do that, so he ran away off the Common. But then he sneaked back on and quickly scurried to a secluded spot where the Brights would not bother him.

Some members of the Bright club were teachers. They taught the children that they should spend their time learning on the Common, and they said that the most important lesson was the lesson of the two rules. Rule 1. Everyone should share their discoveries about the plants and the streams and the stones lying on top of the ground. Rule 2. There should be no digging under the ground or climbing in the trees.

When the children's parents heard about this lesson, some of them liked it, but some complained. The complaining parents said they liked plants and things like that, but they wanted their children to feel free to dig as well, and to learn about minerals and roots. Also, they thought climbing was fun. But the head teacher said, 'What? Don't you want your children to share all the discoveries about the plants and the stream and the stones?' 'Yes, yes, of course we like all that. But we like digging as well.' The head teacher was shocked. 'There is no place for digging in education', he proclaimed.

Soon, whenever anyone went walking on the Common, they would be greeted by people who would say things like, 'Hello! Always nice to meet up with fellow Brights!' If you didn't want to join the Brights then you just had to keep quiet, because whenever anyone mentioned digging or climbing, the Brights would jeer at them and call them names like 'leaf-head'. The Brights would write letters to one another, congratulating themselves on not being ignorant like the other people. 'Don't they care about all the knowledge we have gained?' they would ask. 'Just think of all our careful studies of the water and the plants and the rocks lying on the ground! Those people are just holding us back with all their rooting in the soil. They are trying to push us back into the dark ages!'

One day the village woke to find that big noticeboards had been put up all around the common land, saying 'Brightland: all welcome!' There was a notice explaining about all the things that had been discovered there. The Brights loved these fresh new noticeboards. But when the blacksmith came to read the notice, he said, 'Isn't this the common land, where our parents and grandparents and great-grandparents have been free to roam ever since the foundation of the village?' Some people gathered round, and explained, 'Yes, that's right, the bright common land, isn't it lovely?' 'But I don't want to call it Brightland. It already has a name. It is called the Common.' 'What?' said the Brights, 'Oh, no, it didn't have any name before we discovered it. And now it is called Brightland. Isn't that a lovely name?'

The blacksmith made a move to push the noticeboard over, but a policeman came along and said he could not do that. It would be criminal damage. It would be vandalism. Then the blacksmith started to say insults like 'you thieves!' and 'you snakes!' and worse things, so the policeman took him away. Everyone agreed that he was a dangerous person. A dark-minded person.

Now that the Common was called Brightland, people who were not in the Bright club felt a bit uneasy walking there. They never knew when they might be accused of being ignorant, so they went there less.

After the noticeboards had been up for a few years, they began to look a bit faded, so they were mended and fresh paint was applied. The sign said 'Brightland: all Brights welcome!'

The Brights held big meetings and wrote books. They noticed that they didn't seem to see many non-Bright people walking on the Common (that is, Brightland as they called it). They wondered why this was. After thinking about it for a while, a Bright professor wrote a book to explain it. He explained how Bright people were always eager to learn new things, but the non-Brights were not like that. Those people, he explained, just stayed stuck in their own homes and were not really interested in the study of plants and rocks, water, sunlight and ice.

So it was decided that it would be best to put a high, boarded, wooden fence all around Brightland, to protect the space and to check on who was allowed in or out. It was important to keep everyone safe, after all. The fence had some gates with signs saying

'Brights only'. When people tried to get onto the Common, the guardians of the fence said, 'Brightland belongs to everyone; you may enter just as soon as you have signed the rules'. But some people did not like the rules. When anyone said that, the Brights scoffed and said 'Don't you appreciate the wonders of Brightland? Its great variety of plants, its rich life and all the patterns and the wonderful knowledge to be gained? Your minds must be dark. Dark like the places under the ground. What places are those anyway? They don't even exist!'

CHAPTER 21

GETTING PAST BRIGHTLAND

In this book I have first been positive about the beauties and
wonders and sheer intelligence of scientific insight into the natural world, whether it may be the unfolding of the cosmos from
its earliest moments, or the long, steady process of biological
evolution. I have also been negative—I have proposed that all is
not well in the public presentation of these areas of science. And
I have tried to implement some of the corrective work that needs
to be carried out. But I don't want to leave the discussion on that
negative note. I think the future can be better, and I would like to
suggest how that future might look.

If we don't want the future to be Brightland then we have other
options, and it is time to start enjoying them again. Brightland is
the land of shiny intellects who have expert knowledge of slivers
of science and make up the rest of life as they go along, with
scant patience for hard-won wisdom coming from long-standing
elements of human culture. In Brightland there is an undercurrent of fear, the fear of ridicule, which masquerades as a civilized
determination to be intellectually top-notch. There is a tendency
to view the whole universe as a machine, with ourselves cogs
in that machine. It will be asserted, for example, that a correct
description of the physical world is entirely reductionist. When
someone says that, then ask them: reduced to what? To quantum

Liberating Science. Andrew Steane, Oxford University Press. © Andrew Steane (2023).
DOI: 10.1093/oso/9780198878551.003.0021

fields? But contemporary quantum physics is itself not entirely reductionist, owing to quantum entanglement, and furthermore it is itself a model that is no more likely to be the whole story than was Newtonian mechanics. 'Everything we have learned about neuroscience', someone may say, 'is consistent with the model that the brain is a machine driven by physical laws of the type we already see at work in physics and chemistry.' That is like someone who, having waded out a hundred yards into the sea off West Africa, then declares, 'Look: the sea is not getting much deeper. Everything I have discovered so far is consistent with the claim that I can walk in these shallows all the way from here to South America.' The statement is a true one, but entirely misleading.

There is also a tendency to swallow ideas such as the idea that free will is an illusion, in the belief that this is what scientifically grown-up people ought to think.

We can begin to remove the Brightland fence, and return the common land to its ancient freedom, by invoking the main conclusions of the first and second halves of this book. The main conclusion of the first half is that astrophysics and quantum physics do tell us many deep things about the history of the universe as a whole, and about what physical stuff is like in its innate nature, but our ultimate origins, whether in ancient times or in the roots of reality in the present moment, remain beautifully mysterious. By 'mysterious' I don't mean 'don't ask questions'; I mean on the contrary: *do not think that you or anyone else have already got it pinned down, so do ask questions.* In particular, I suggest, be sceptical of the claim 'Oh well, it just happened', when it comes to our origins, or at least understand that that kind of appeal is not a scientific one. Also, realize that no area of study can fully account for its own terms and assumptions. Quantum physics will not tell you our origins, because quantum physics cannot explain quantum physics. And anyway quantum physics is not the last word on the physical nature of anything. The complete truth about the physical nature of anything can never be certainly known by us, and it is quite likely more subtle than any model we have yet thought of.

Bad religion cannot tell you our origins either, because bad religion is muddled and self-absorbed. Good religion cannot settle it either, because good religion invites us to be entirely honest about our lack of final knowledge and redirects us to an attitude of seeking, grieving, serving and celebrating. As we accept this lack of final knowledge, combined with the invitation to seek, we find ourselves living in a more liberating way, one which is welcoming towards science without being either ground down by science or over-inflated by it.

So that was a reflection on the first half of this book.

In the second half of the book I have aimed to get in view a more coherent, balanced and well-reasoned view of what does and does not follow from evolutionary processes in biology. The main conclusion of this part is that the biosphere is replete with pattern-forming processes, and what we inherit from our evolutionary history is not a set of functions which we helplessly fulfil like so many gene-replicating robots, but rather a set of capacities and abilities to explore. These include capacities which allow us to negotiate the physical environment and also capacities which allow us to discover and learn the abstract environments of logic, aesthetics and ethics. We find that, far from being mere slaves to impersonal forces beyond our control, we are, to a very significant degree, free to exercise our gifts and responsibilities as moral beings. That is what has been conferred on life on Earth, and possibly life on other planets in the cosmos, by means of the processes that have been enacted and embodied. So let's celebrate this aspect of our evolutionary history!

The heart of the book may be found in the footnote towards the end of Chapter 14. It is the scene of myself as an older person, in discussion, during a short car journey, with a younger person, someone on the verge of adulthood and working out their values and understanding. I feel that the idea I passed on in that conversation—the idea that values are not arbitrary, and we can perceive them, and Darwinian evolution does not suggest otherwise—is a great treasure. I think many people nowadays either believe in value and reject significant parts of science, or accept science and reject the objectivity of value—or else they try to hold to both but experience a cognitive dissonance because they think they are in contradiction. How liberating it is to be freed from that dissonance and embrace the riches of our true inheritance!

Evolution has been presented in many discussions as, in essence, a fight or a sequence of survival problems, like a perpetual end-of-year examination. It does, I think, include those aspects, but they are far from the whole story, and they are not the central story. Evolution is also a dance. A dance of possibility and of discovery. Not so much an examination as an education. Here is an apt remark of McGilchrist's (by 'life' here he means life in all its abundant forms and processes, not just human life):

> Life, in its essence, is a making new: a wholly superfluous, super-abundant, self-overflowing—an exuberant, self-delighting process of differentiation into ever more astonishing forms, an unending dance, in which we are lucky enough to find ourselves caught up—not just, as the left hemisphere cannot help but see it, a series of survival problems to conquer.
>
> I. McGilchrist, *The Matter with Things*, vol. 2, p. 853.

Turning now to the later chapters of this book, I have emphasized that capacities such as reason, science and rationality are human universals, and to contrast them with a thoughtfully grounded religious commitment, as if the one could only exclude the other, is both unjust and irrational, like claiming that men are more reasonable than women, or that algebra is an alternative to mathematics. We need to be wary of the mindset that says 'I and my group are the true champions of reason' or 'reason is on our side' before some argument even commences. Instead, we must take up the patient task of *showing* that reason is on our side of some argument, if indeed it is.

Bringing together now both the remarkably structured nature of our physical origins, and our inherited capacity to discover and explore various types of 'world' (physical, rational, aesthetic, moral), we are able, if we choose, to put all this into practice by being people who can love and be loved.

Love is free gift, or it is not love at all. Which means we need to know whether it is correct to say, of ourselves, that we have some freedom of choice. A further important contribution which we can make to removing the Brightland fence is to reassert the notion of free will. There is nothing wrong with free will, and a lot right with it. Psychology and neuroscience have, admittedly, made discoveries which show our freedom of thought and decision-making is not as complete as was once thought, but this is far from undermining it completely. We do have large amounts of freedom to decide how to occupy our minds—what to pay conscious attention to, and what to turn from. At least, it is perfectly reasonable so to think, because there is plenty of evidence of such freedom in human behaviour, and more generally the sensitivities in many physical systems make it impossible to tell, from the physical evidence, whether determinism holds in the world at large. What this implies is that our intuition that we and others have some genuine freedom of choice does not need to be rejected as delusion, but can be embraced as a good insight—indeed, one of the best we have. Neither physics nor neuroscience is competent to overturn that reasonable position.

It may be that when we are forced to make rapid decisions about inconsequential matters, our response is a reflection of a way of thinking that has become automatic for us. In this way, some decisions may not be free in the moment we think we take them, but they are the outcome of a freedom we had in the shaping of our own thoughts over a long period. A fitting encouragement in this direction can be found in a passage which was read out to me and my fellow pupils at the end of each year at school, as follows:

> Finally, brethren, whatsoever things are true, whatsoever things are honest, whatsoever things are just, whatsoever things are pure, whatsoever things are lovely, whatsoever things are of good report; if there be any virtue, and if there be any praise, think on these things.
>
> From the letter of St Paul to the church at Philippi,
> translation, King James Version.

This reminds us that the choice to *think on these things*—to direct our mental gaze one way as opposed to another—can have a powerful effect on us. This is a large component of human freedom.

Another component of human freedom is the freedom *to decide how to respond* to the situation we find ourselves in. This aspect was emphasized by the neurologist, psychiatrist and philosopher Victor Frankl. Our situation may be unjust, constraining and controlling, as Frankl knew well from his experience in Nazi concentration camps. His considered view was nonetheless that the responses of the prisoners were varied and capable of asserting the meaning of their lives even in the face of the inhuman controls placed on them. In this extreme of deprivation, the freedom to formulate one's own thoughts and judgements and attitudes was perhaps the last dignity which no captor could take away. We should certainly resist the claim that such a dignity is illusory. The claim that we have no freedom should be required to show credentials of the utmost quality before it deserves to be accepted, and it is nowhere near having such credentials on the scientific knowledge accumulated so far.

Having recovered the freedom to assert free will, and the rational powers that go with it, we can also return to an appreciation of, and full enjoyment of, all the range of human intellectual activity throughout the arts and humanities as well as the sciences. We do not need to regard all these endeavours as second-class citizens in the intellectual landscape, waiting for science to come along and tell them how to behave. We can recognize that the various disciplines are exploring landscapes of different kinds—landscapes of ethics and aesthetics and history and philosophy; landscapes of brotherhood and of parenthood and of the pains of loss; landscapes of the human heart, and of any heart that is able to love and to suffer. These various landscapes are explored by the living of life as fully and sensitively as we can, and they are illuminated by the exchange of ideas, and by developing languages adequate to express them. Quantum field theory will not provide a better language.

Enjoying and appreciating the common land, we will come to understand, as I expressed it in *Science and Humanity*, that

> science is about building up an insightful picture in which the underlying microscopic dynamics do not replace, nor do they

explain, the most significant larger principles, but rather they give examples of how those larger principles come to be physically embodied in particular cases. The lower-level and higher-level principles are in a reciprocal relationship of mutual consistency in which each illuminates the other.

Children will learn this at school, and it will encourage them both to see science truthfully and insightfully, and to embrace what it offers. It will enable them to respect the scientific effort, and join in with it, without succumbing to the dehumanizing tendency which springs up when it is turned into an idol or an all-purpose oracle, supposed to be able to declare every truth worth hearing. As these children become adults, it will be harder to sell to them grotesquely unfair ideas on the grounds that they are said to be 'scientific'. Equally, they will be receptive to the warnings and guidance which scientific study offers to public policy, because they will not fear that some other agenda is being pushed. Teachers will feel that they are partners in an effort which is part of a rounded human life, not an attempt to replace hard-won wisdom and moral insight with a project of measurement and manipulation towards some political end. Religious organizations will be enthusiastic champions of science—science the genuine article, as it is practised in institutions with a good record and open doors. Universities will have generous instincts of support for the best aspects of religion, and it will no longer be clever for people who have never read any major work of theology to use that word as a joke.

Once the common land is free again, there will be no ghettos.

Hope

Here is my lovely sign of hope.
It will sound impossible, but bear with me.
It is everywhere, but you will not find it
by looking at any particular place.
It is far from you but also close,
as close as you are to yourself.
It is not on the map, and you will
not find it by digging, nor by sending
up balloons to assay the sky.
It is nowhere but everywhere, yet
in each place you can find the whole of it.
All this, because it is not a location
but a direction. It is not a place you
can claim but a way to look.

You must not hide your shadow behind you
but throw it down before. Then,
stay still, focus on infinity,
and look in a particular direction.
And suddenly there it is,
layer on layer in all its colours
as luminous as if it were refreshed
just that instant, replenished
again with fullness of arching light
breaking out of the water.

INDEX